Springer Series in Advanced Manufacturing

The **Springer Series in Advanced Manufacturing** includes advanced textbooks, research monographs, edited works and conference proceedings covering all major subjects in the field of advanced manufacturing.
The following is a non-exclusive list of subjects relevant to the series:

1. Manufacturing processes and operations (material processing; assembly; test and inspection; packaging and shipping).
2. Manufacturing product and process design (product design; product data management; product development; manufacturing system planning).
3. Enterprise management (product life cycle management; production planning and control; quality management).

Emphasis will be placed on novel material of topical interest (for example, books on nanomanufacturing) as well as new treatments of more traditional areas.

As advanced manufacturing usually involves extensive use of information and communication technology (ICT), books dealing with advanced ICT tools for advanced manufacturing are also of interest to the Series.

Springer and Professor Pham welcome book ideas from authors. Potential authors who wish to submit a book proposal should contact Anthony Doyle, Executive Editor, Springer, e-mail: anthony.doyle@springer.com.

More information about this series at https://link.springer.com/bookseries/7113

Lei Deng · Juchen Xia · Xinyun Wang

Precision Forging Technology and Equipment for Aluminum Alloy

国防工業出版社
National Defense Industry Press

Springer

Lei Deng
School of Materials Science
and Engineering
Huazhong University of Science
and Technology
Wuhan, Hubei, China

Juchen Xia
School of Materials Science
and Engineering
Huazhong University of Science
and Technology
Wuhan, Hubei, China

Xinyun Wang
School of Materials Science
and Engineering
Huazhong University of Science
and Technology
Wuhan, Hubei, China

The book is funded by B&R Book Program.

ISSN 1860-5168 ISSN 2196-1735 (electronic)
Springer Series in Advanced Manufacturing
ISBN 978-981-19-1827-8 ISBN 978-981-19-1828-5 (eBook)
https://doi.org/10.1007/978-981-19-1828-5

Jointly published with National Defense Industry Press
The print edition is not for sale in China (Mainland). Customers from China (Mainland) please order the print book from: National Defense Industry Press.

This Springer imprint is published by the registered company Springer Nature Singapore Pte Ltd.
The registered company address is: 152 Beach Road, #21-01/04 Gateway East, Singapore 189721, Singapore

Preface

As the preferred lightweight metal material, the application of aluminum alloy has expanded from the aerospace field to civil industry and conventional weapons fields to manufacture lightweight products, such as automobiles, high-speed trains, motorcycles, high-speed ships, tanks, armored vehicles, and guns.

In the early 1990s, industrialized countries have begun to develop and apply lightweight technology for aluminum alloy automobile bodies and related parts. At present, in the United States, Germany, and Japan, the amount of aluminum alloy used in an automobile has reached 20% of its own weight. However, with the continuous improvement of emission standards and the Government's emphasis on environmental protection in the worldwide, the research and development of aluminum alloy parts still needs to be further enhanced, and the demand for aluminum alloy precision forging technology has been paid attention to.

The author of this book began to study the precision forging technology of aluminum alloy in the 1990s, as well as precision forging equipment that meet the needs of precision forging. A number of theoretical and application achievements have been obtained. Based on these achievements and the latest research results of aluminum alloy precision forging technology, the book is edited for the reference of scholars in the field of plastic forming of aluminum alloy.

The book is co-edited by Lei Deng, Juchen Xia, and Xinyun Wang of Huazhong University of Science and Technology. Junsong Jin and Peng Zhou of Huazhong University of Science and Technology are responsible for the compilation of Chaps. 7 and 6 respectively. In addition, Yunjun Zhang, Tianfu Chen and Yang Yan of Hubei Tri-ring Forging Co., Ltd. participated in the writing of precision forging technology for long shaft parts in Chap. 4, and Yi Feng, Jun Yu and Zili Xia of Wuhan Newish Technology Co., Ltd. participated in the writing of servo hydraulic press and electric screw presses in Chap. 7.

Thanks to the previous students of the Precision Plastic Forming Group of the State Key Laboratory of Material Processing and Die&Mould Technology, Huazhong University of Science and Technology for their work, including Junwei Cheng, Yarui Zhang, DongSheng Ji, Qingjie Li, Ting Zhao, Zhihua Ren, and Ruiqi Xiao for their outstanding contributions to the content of this book, Haidong Zhang, Fei Du, Cheng

Li, Fanyi Zeng, Zhuang Tian, Hao Zhang, and Ming Zhu for their assistance in proofreading the book.

Thanks to the National Natural Science Foundation of China (Project Nos.: 52090043, 51725504, 51675200, 51205143) for funding related research work. This book has also been funded by the Silk Road Book Fragrance Project of the National Press and Publication Administration of China.

Wuhan, China

Lei Deng
Juchen Xia
Xinyun Wang

About This Book

The book is divided into 7 chapters. The application of aluminum alloys and the research trends of precision forging technology for aluminum alloy are summarized in Chap. 1. Chapter 2 introduces the foundation of aluminum alloy precision forging technology. Chapter 3 presents the finite element simulation technology of aluminum alloy precision forging. The precision forging forming process and die design of typical aluminum alloy parts including long shaft, complex revolving body, and branch-shaped parts are discussed in Chaps. 4, 5 and 6, respectively. Chapter 7 introduces the development and application of aluminum alloy precision forging equipment.

The book has the following four characteristics:

(1) Taking precision technology as the main line and combined with die design and equipment development, a complete knowledge system of aluminum alloy precision forging technology and equipment is presented.

(2) In addition to conventional precision forging technologies and equipment, the new technologies and new equipment including closed precision forging, flow control forming, squeeze casting forming, combined casting-forging forming, new precision forging hydraulic press, servo hydraulic press are discussed in detail, which will help inspire readers to further develop new technologies and new equipment.

(3) When introducing examples of precision forging for typical parts, the process parameters calculation, numerical simulation analysis, die design, and process test are presented, which will help the readers to understand more deeply and comprehensively.

(4) The book can be used as a reference book for those who are engaged in the fundamental research and application development of precision forging technology for aluminum alloy, other light alloys, and ferrous metal in the field of plastic processing.

Contents

Chapter 1
Introduction

1.1 Application of Aluminum Alloy Forgings

With the continuous increased requirements in automobile lightweight and environmental protection, the application fields of aluminum alloy forgings has become wider and wider. It has been extended from the aerospace field in the past to many civil industrial fields, such as automobiles, high-speed trains, motorcycles, high-speed ships, weapons, construction, etc., in which the use of aluminum alloy forgings are increasing [1, 2].

1.1.1 The Field of Transportation Vehicles

The application of aluminum alloy in the automotive field has been continuously expanding whether it is from the types of parts or the proportions. Aluminum alloy has many advantages as an automobile material. For example, the weight of aluminum alloy forgings is about 60% of that of steel under the condition of meeting the same mechanical properties. The aluminum alloy can absorb 50% more energy than steel in crash tests. Moreover, anti-rust treatment is not required for aluminum alloy [3]. Because of these outstanding advantages, its proportion in automotive materials is increasing year by year. Take the United States as an example, in the 1980s, the average aluminum alloy used in an automobile was 55 kg, by the 1990s it had reached 130 kg, and now it has exceeded 200 kg which is more than 18% of the total weight of automobiles [4].

Figure 1.1 shows the amount of aluminum alloy used in automobiles of several countries in 2010 [5]. It can be seen that China's total consumption of aluminum alloy in automobiles in 2010 was 2.32 million tons, which is the most in the world. However, the amount of aluminum alloy used in an automobile in China is only 127 kg, which is 12% lower than that of the United States and 15% lower than that

L. Deng et al., *Precision Forging Technology and Equipment for Aluminum Alloy*, Springer Series in Advanced Manufacturing,
https://doi.org/10.1007/978-981-19-1828-5_1

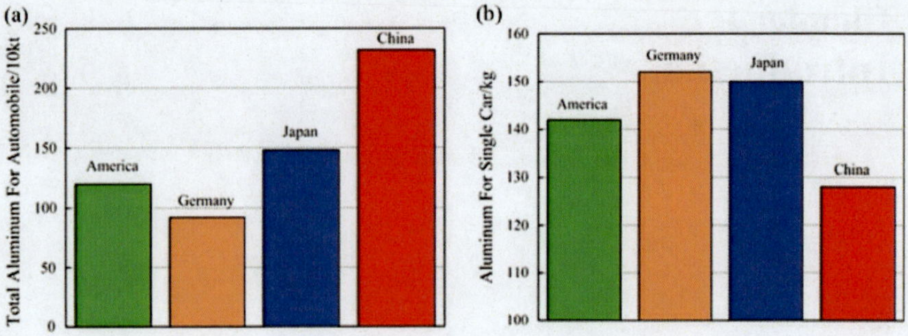

Fig. 1.1 Aluminum alloy consumption in automobiles of several countries. **a** The total amount of aluminum alloy used in automobiles. **b** The amount of aluminum alloy used in an automobile

of Japan and Germany. With the further development of the automotive industry, the application prospects of aluminum alloys in automobiles will become increasingly broad, especially in developing countries.

General Motors Company used 7021 aluminum alloy plate to manufacture the bumper reinforcement bracket of Saturn car, and Ford Motor Company also used this material to manufacture the bumper reinforcement bracket of Lincoln Town car. At present, not only the piston, radiator, and oil pan of engine are made of aluminum alloy, but also the cylinder head and crankcase are also made of aluminum alloy. The high-precision aluminum alloy cylinder head made by Roves is only 36% of the weight of the cast iron cylinder head. The LuPo3LTDI automobile of Volkswagen used a variety of aluminum alloy, making its total weight only 830 kg, which is 230 kg lighter than other LuPo series automobiles. The entire frame of the Audi A8 is made of aluminum alloy, and the covers are stamped from aluminum alloy sheet [6]. Compared with the steel body, this kind of aluminum alloy body can reduce the weight by 30–50% and reduce the fuel consumption by 5–8%. The aluminum alloy used for the body of the Insight hybrid automobile produced by Honda is 162 kg, which is about 40% lighter than the steel body. The front axle tie rods, lateral guide arms, and the overall support structure of the front axle of Mercedes-Benz's S series automobiles are also made of aluminum alloy. The weight of these components is only 10.5 kg, which is 35% of the steel [7].

In terms of aluminum alloy forgings, due to their high price, they have not been widely used in automobiles. They are only used in small amounts on automobiles of Europe and the United States. The weight of aluminum alloy forgings used on each automobile is about 40 kg, which only accounts for about 20% of aluminum alloy used in automobiles. However, because forged aluminum alloy has the advantages of high specific strength and smooth surface, its application in automobiles is gradually increasing. At present, aluminum alloy forgings have been used in the suspension system of some automobiles, such as Mercedes-Benz S-Class and Honda NSX. Some typical aluminum alloy forgings for automobiles are shown in Fig. 1.2. They are mainly used for the chassis components of automobiles and light vehicles, such as wheels, drive shafts, suspension parts, control arms, left and right tie rods, tripods,

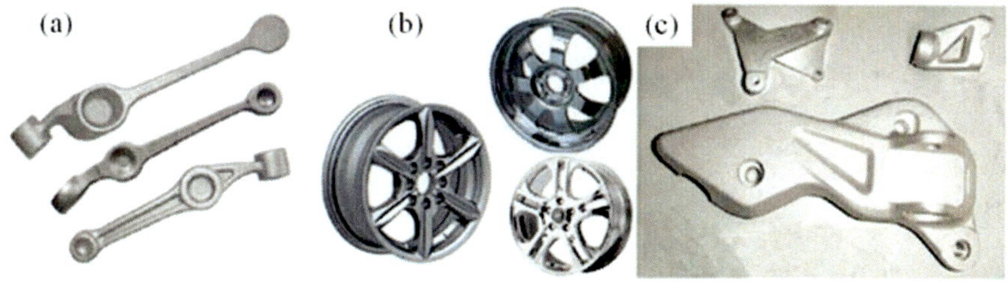

Fig. 1.2 Some aluminum alloy forgings: **a** control arm, **b** wheel, **c** support plate

steering knuckles, etc., and now are beginning to be used for engine parts, such as connecting rods and pistons [8].

Exhaust emission of automobiles has become a major source of air pollution and accounts for about 10% of the pollution sources. About 60% of the energy consumption of an automobile is consumed by its own weight. The weight of the automobiles is reduced by 100 kg, its fuel consumption can be reduced by 0.4 L, and exhaust emission can be reduced by 1 kg for every 100 km. In 2016, China's auto production and sales were 28.03 million, ranking first in the world. The Society of Automotive Engineers of China predicted that China's automobiles will continue to grow. Lightweight materials and their manufacturing technology have become an important way to achieve energy saving and emission reduction [9, 10].

In addition, the fast-developing high-speed trains in recent years not only use aluminum alloy to make some internal parts, but also the train body including the frame, bottom plate, control components and body panels, are made of aluminum alloy. The total length of China's high-speed railway will reach 30,000 km, and the required trains will exceed 8,000. Therefore, the required amount of aluminum alloy parts is huge.

1.1.2 The Field of Military Weapons

In the field of military weapons, more aluminum alloy parts are used in armored vehicles, tanks, torpedo boats, howitzers, firearms and other weapons than before. For example, 105 mm caliber howitzers are made of aluminum alloy parts including large frame, cradle, front seat plate, left and right trunnion brackets, and aiming mirror bracket, drawbar, balancer outer tube, etc. After using aluminum alloy, the weight of the gun was reduced from 3.7 tons to 1.4 tons. For another example, wrought aluminum alloys are used to manufacture armored car body, connecting rod base, brake disc, steering knuckle, track tensioner, inducer, road wheel, turret seat, smoke launcher, ammunition rack, storage compartment, fuel tank, seats, pipes of tanks, and cast aluminum alloys are used for tank diesel engine cylinder heads, cylinder blocks, crankcases, pistons, compressor impellers, compressor casings, booster boxes and drive shafts. In firearms, the upper and lower receiver body, the gun frame buffer seat, the firing base, the gun tail closed receiver, the tripod, the grenade shell, the grenade launcher (barrel), and the rocket are made of high-strength aluminum alloy.

1.1.3 Other Fields

1. Construction

In the construction industry, aluminum alloys are mainly used in the framework, roofs, wall enclosures, frames, doors and windows, suspended ceilings, veneers, floors, sun-shading components and decorations of public facilities, industrial facilities, agricultural facilities and buildings; storage tanks for acid, alkali and various liquid and gaseous fuels, inner walls and conveying pipes of reservoirs; straddle structures, barriers and bridges of highways, pedestrian passages and railways; scaffolding, pedals and elevators.

2. Packaging

At present, the main application forms of aluminum alloy in the container packaging industry include rigid all-aluminum cans, boxes, bottles, pots, barrels, pots; semi-rigid boxes, cups, cans, shallow plates, plates and other flexible packaging; household foil, food packaging foil; sealing sheets and caps of various bottle parts; hoses and other products packaging, etc.

3. Electrical engineering

Aluminum alloy has good electrical conductivity, so it is widely used as a conductor in electrical equipment. It is mainly used in the electric power and telecommunications industries in the form of wires, pipes, and foils. It is also used to manufacture various electrical appliance casings and covers in the form of plates. Aluminum alloy is also used for heat sinks of high-power electrical components in various shapes and structures.

4. Petrochemical equipment

The petrochemical equipment made of aluminum alloy includes containers, towers, heat exchangers, and various pipelines. In the 1960s, the United States used 2014 aluminum alloy pipes in oil and gas drilling projects. Aluminum alloy pipes instead of steel pipes can increase the drilling capacity of the rock drill by 50–100%, and save fuel consumption by 15–20%. The total length of each piece of equipment can be increased by 60%. The application of aluminum alloy drill pipe has greatly increased the drilling depth, which is of great significance for drilling deep and ultra-deep wells. In addition, aluminum alloy drill pipes will not cause sparks during work, thus ensuring the safety of oil and gas development.

1.2 Research on Aluminum Alloy Precision Forging Technology

Aluminum alloy is the preferred lightweight material for the manufacture of the parts of aerospace vehicles, artillery and conventional weapons, automobiles and high-speed trains. Due to the advantages of high product precision, good mechanical properties, and small machining allowance, aluminum alloy precision forging technology has attracted great attention from researchers. It has developed rapidly and achieved many innovative results. Its research progress can be summarized into the following four aspects.

1.2.1 The Demand for Lightweight Manufacturing

In the early 1990s, global automobile production has remained above 80 million vehicles. The lightweight of automobiles is of great significance to energy conservation, emission reduction and environmental protection. The top ten automobile manufacturing companies in the United States, Europe, Japan and other countries jointly established an automobile body lightweight technology research institution, which specializes in the planning and implementation of automobile lightweight technology. Honda Motor issued a strategic report in October 2004, and plans to use 200 kg of aluminum alloy parts on automobiles, of which 40 kg are forgings.

According to the survey of China's passenger automobile samples, the weight of electric vehicles is reduced by 10%, the driving distance can be increased by more than 5.5%. It can be seen that the application prospect of aluminum alloy in new energy vehicles is also very broad. Take the tank truck as an example, if a 45,000 L tank truck is made of aluminum alloy, the weight can be reduced by about 2,500 kg compared with the traditional steel tank truck. In terms of fuel consumption, for every 100 kg of weight reduction, it can save 0.4–0.5 L of fuel consumption per 100 km. The fuel consumption is reduced by 1L, the CO_2 emissions can be reduced by 2.33 kg. Thus, if the tanks of the tank trucks are all made of aluminum alloy, the annual CO_2 emission can be 21.5 million tons.

The lightweight of chassis parts can reduce fuel consumption by more than 40% than body parts. Meanwhile, it enables the automobile to have better dynamic response and handling, thereby improving the comfort and safety of the vehicle. The control arms and steering knuckles of various independent suspensions used in modern automobiles, such as wishbone, trailing arm, MacPherson and multi-link suspensions, can be made of aluminum alloy. In addition to suspension parts, aluminum alloy can also be used to make wheels and other parts. For the manufacture of these parts, precision forging is a potentially advanced technology, which drives the rapid development of aluminum alloy precision forging technology.

1.2.2 Flow Control Forming Technology

Flow Control Forming was proposed by German and Japanese scholars after the mid-90s of the last century. It is a new closed precision forging technology developed on the basis of conventional closed die forging. For complex and difficult-to-form forgings, the billet is placed in a closed die. When the punch exerts a forming force, a strong three-direction compressive stress is generated in the metal. The plastic forming ability of the material is improved due to the increase of hydrostatic pressure. And through the reasonable configuration of flow control methods, a compressive stress gradient field with an absolute value ranging from small to large is formed from the most difficult filled part to the entrance of the die cavity to ensure that the most difficult filled part and other parts are filled simultaneously. Because the precision forging is achieved by controlling the flow of metal, so it is called flow control forming. Flow control forming is not only suitable for the precision forming of ferrous metal parts such as spur bevel gears, but also for the precision forming of various aluminum alloys, especially high-strength aluminum alloys and other difficult-to-deform metals.

1. Characteristics of flow control forming technology

    (1) The plastic flow of the metal can be accurately controlled, and its forming performance can be improved to realize precision forming of complex forgings.
    (2) The occurrence of defects such as folding and underfilling can be effectively avoided. The streamline of forgings is continuous and dense, which improves the mechanical properties of the product.
    (3) More smooth surface and higher dimensional accuracy of forgings can be obtained. The tolerance level can reach IT8–IT9, which is one level higher than the dimensional tolerance level of general extrusion parts.

2. Methods to realize the control of metal flow

    (1) Applying damping force. Through the reasonable configuration of the reverse force that is the damping force and the forward forging force, a compressive stress gradient field is formed in the die cavity from the entrance to the most difficult filled part.
    (2) Relieving local pressure. By setting the relief cavity in the most difficult filled part of the die cavity, the surface of the forging part at the relief cavity is a free surface, thereby a stress gradient field from the entrance to the most difficult filled part can be obtained.
    (3) Combination of damping and relieving. It is mainly used in the case of precision forging with two steps of closed pre-forging and final forging.

3. Research and development trends

    According to the structural characteristics of the aluminum alloy scroll that the shape is spiral and the wall thickness gradually decreases from the center to the edge, the damping force and the forward forging force are used together to force the metal to flow from the center to the edge. Throughout the forming process, the front end is always flush to ensure precision forming. If the damping force is not applied, the defect of underfilling will occur at the front end of the forging. For the flow control forming process of aluminum alloy scrolls, the Beijing Research Institute of Mechanical and Electrical Technology studied the reasonable configuration between the damping force and the forward forging force. According to the structure characteristics of the cover and shell of airbag gas generator, Huazhong University of Science and Technology proposed the relieving flow control technology for multi-layer thin-walled cylinders, and studied the effect of deformation mode, the position and dimension design of the relieving cavity on the forming quality of forgings. Taizhou Keda Precision Forging Co., Ltd. designed and manufactured two kinds of flow control forming dies, and successfully realized the production of precision forgings of scroll, cover and shell with various specifications.

1.2.3 Combined Casting-Forging Forming Technology

For medium and high silicon aluminum alloy parts, the traditional forming method is mainly to adopt squeeze casting technology which is divided into indirect squeeze casting and direct squeeze casting. The unit pressure during indirect squeeze casting is generally 60–100 MPa. The unit pressure during direct squeeze casting is generally 25–50 MPa. The pressing velocity is 0.2–0.4 mm/s for small aluminum castings and 0.1 mm/s for large aluminum castings. Due to the above process characteristics, the aluminum alloy parts fabricated by these two processes have low density and low strength, and are not suitable for parts that bear large loads, especially dynamic loads. In 2003, Huazhong University of Science and Technology studied precision forging process instead of die casting process, and developed relieving pre-forging and flat thin flash final forging process based on a bar stock fabricated by squeeze casting. The mass production of aluminum alloy piston tail, piston body and swash plate were realized by this process.

In recent years, for complex steering knuckles made of high-silicon aluminum alloy, combined casting-forging precision forming technology has been studied. The general process route is: ingot melting → squeeze casting → trimming and removal of sprue riser → heating → small flash precision forging → trimming. This new process combines the advantages of squeeze casting and closed die forging, and has good application prospects.

1.2.4 Development of New Precision Forging Equipment

Traditional die forging equipment for aluminum alloy parts is die forging equipment for ferrous metal, such as die forging hammers, friction presses, hot die forging presses and hydraulic presses. The forging velocity of the die forging hammer is 5–6 m/s. For high-strength aluminum alloys with high rate sensitivity, it is easy to generate cracks during die forging. Therefore, the die forging hammer can only be used for the open die forging of aluminum alloys with low strength and good plasticity, but this will result in low utilization of materials. The loading characteristics of the friction press is similar to the die forging hammer, so there are similar problems. The hot die forging press is the main die forging equipment currently used for aluminum alloys, but it is mainly suitable for mass production. It is characterized by high efficiency, but low material utilization. For large-scale aluminum alloy rib-web parts of aircraft, large-tonnage die forging hydraulic presses were designed and manufactured for die forging, especially in China an 800MN die forging hydraulic press with the largest forging force in the world was developed. For small and medium-sized high-rib parts, the isothermal precision forging hydraulic press was designed and manufactured to solve the forming problem, but the service life of the die was very low.

Aiming at the problem that the existing die forging equipment cannot adapt to the properties of various aluminum alloys and the structural characteristics of different parts, Huazhong University of Science and Technology has cooperated with Wuhan Newish Technology Co., Ltd. and Hubei Tri-ring Metalforming Equipment Co., Ltd. for more than 10 years, and developed the J58K electric screw presses, YK34J double-acting hydraulic press and YK34J multi-directional die forging hydraulic press, which meet the requirements of precision production of aluminum alloy forgings.

References

1. Liu B, Peng CQ, Wang RC, Wang XF, Li TT (2010) Recent development and prospects for giant plane aluminum alloys. Chin J Nonferrous Metals 20(9):1705–1715
2. Liu JA (2005) Current situation of aluminum forging production and application prospect of forgings. Aluminium Fabr 2(161):5–9
3. Wang LA (1994) Forging technology and its developing in aeronautical industry. Forg Stamp Technol 1994(1):57–61
4. Tian FQ, Li NK, Cui JZ (2005) Research and development of ultra high strength aluminum alloys. Light Alloy Fabr Technol 33(12):1–9
5. Wang ZT, Zhang XH (2011) Aluminum alloy for automobile usage. Light Alloy Fabr Technol 39(2):1–14
6. Hua L (1999) New materials and advanced manufacturing technology for automobiles. Automob Technol Mater 4:7–9
7. Feng MB (2006) Development and applications of new materials in automotive lightweighting technologies. Automot Eng 28(3):4–11

8. Wei W (2013) Research on hot deformation behavior and microstructure property of 6082 aluminum alloy forging with rib. China Academy of Machinery Science and Technology, Beijing
9. Huang PX (2002) Application of aluminum alloy in automobile. Shanghai Auto 1:37–38
10. Ma MT, Ma LL (2008) Application and prospective technology of aluminum alloy in automobile lightweight. Adv Mater Ind 9:43–50

Chapter 2
Fundamental of Precision Forging Technology for Aluminum Alloy

2.1 Overview

The concept of aluminum alloy precision forging technology is the same as that of ferrous metal precision forging technology that is the shape and dimensional accuracy of the forged parts are as close as possible to, or even the same as, the shape and dimensional accuracy of used parts. The mechanical properties of the forged parts also meet the requirements.

Compared with the traditional die forging process, precision forged parts have small machining allowance, dimensional tolerance, and die forging slope, therefore, the material utilization is high, and the subsequent machining time is reduced. Along with these advantages, the effects of energy saving and environmental protection will been produced.

According to the performance characteristics of aluminum alloys and the structural characteristics of parts, the precision forging process mainly has the following three methods.

(1) Small flash precision forging process. It is also known as the flat-thin-flash precision forging. The flash around the forging is a very thin structure which only has the bridge part and without warehouse part. Compared with the traditional die forging process, its flash volume can be reduced by more than 60%. This process method is suitable for the precision forging of long shafts, and complex rod and plate parts with large horizontal projection area and small thickness.
(2) Non-flash closed precision forging process. The aluminum alloy forgings produced by this process do not generate lateral flash around the parts, only have small longitudinal burrs. This process is suitable for the precision forming of revolving bodies and non-revolving parts with simple outline, especially the high-strength aluminum alloy parts.

L. Deng et al., *Precision Forging Technology and Equipment for Aluminum Alloy*, Springer Series in Advanced Manufacturing,
https://doi.org/10.1007/978-981-19-1828-5_2

(3) Combined casting-forging process. For complex branch-shaped and fork-shaped aluminum alloy parts, squeeze casting process can be used to the preform of the part, and then small flash die forging is used to form the final precision part.

Although three different precision forging processes have different characteristics, the basis of the precision forging process is similar. This chapter has carried out a more comprehensive and systematic elaboration about the theoretical foundation and key technologies of aluminum alloy precision forging.

2.2 Types of Aluminum Alloys

Aluminum alloys can be divided into two major categories: wrought aluminum alloys and cast aluminum alloys. Wrought aluminum alloys are further divided into two subcategories: heat-treated strengthening and non heat-treated strengthening. According to different chemical compositions, these two subcategories are divided into four series respectively. Cast aluminum alloys are also divided into two subcategories according to whether it can be strengthened by heat treatment. See Table 2.1 for more details.

The forgeability of ten commonly used aluminum alloys is shown in Fig. 2.1. It can be seen that the forgeability of aluminum alloys increases with the increase of temperature, but the forgeability has different degrees of sensitivity to temperature. For example, the forgeability of 4032 aluminum alloy with high silicon content is very sensitive to temperature changes, while high-strength 7075 aluminum alloy is less affected by temperature. The reason is that the types and contents of alloying elements are different, and the quantity and distribution of strengthening phases are also different, which affects the plasticity and resistance to deformation of aluminum alloys.

Another property related to the forgeability of aluminum alloy is plastic fluidity which mainly depends on the deformation resistance and friction factor of the alloy. The smaller the deformation resistance and friction factor, the better the fluidity. At the forging temperature, the deformation resistance of high-strength aluminum alloy is greater than that of steel, and the friction factor is also larger, so the fluidity of high-strength aluminum alloy is poor.

It can be seen from Fig. 2.1 that the forging performance of 6XXX series aluminum alloy is the best, the forging performance of 2XXX series aluminum alloy is in the middle, and the forging performance of 5083 and 7XXX aluminum alloys is poor. In order to quantitatively compare the forgeability of aluminum alloys, the amount of deformation generated under a certain energy is defined to characterize the forging performance, as shown in Fig. 2.2. In addition, the forging ratio of aluminum alloy should also be paid attention to. In order to ensure the stability during forging, the forging ratio of aluminum alloy should be designed within an appropriate range.

Table 2.1 Classification of aluminum alloy

Category	Subcategory	Series
Wrought aluminum alloy	Heat-treated strengthening aluminum alloy	Al-Cu series alloy—2XXX series, such as 2024 alloy Al–Mg–Si series alloy—6XXX series, such as 6063 alloy Al–Zn–Mg–Cu series alloy—7XXX series, such as 7075 alloy Al–Li series alloy—8XXX series, such as 8089 alloy
	Non heat-treated strengthening aluminum alloy	Pure aluminum—1XXX series, such as 1050 alloy Al–Mn series alloy—3XXX series, such as 3004 alloy Al–Si series alloy—4XXX series, such as 4032 alloy Al–Mg series alloy—5XXX series, such as 5083 alloy
Cast aluminum alloy	Heat-treated strengthening aluminum alloy	Al–Mg–Si series alloy, such as ZL107 alloy Al-Cu–Mg–Si series alloy, such as ZL110 alloy Al–Mg–Si series alloy, such as ZL104 alloy Al–Zn–Mg series alloy, such as ZL402 alloy Al–Zn–Si series alloy, such as ZL401 alloy
	Non heat-treated strengthening aluminum alloy	Pure aluminum Al–Si series alloy, such as ZL102 alloy Al–Mg series alloy, such as ZL301 alloy

Figure 2.3 shows the relationship between forging ratio and mechanical properties of 2014-T6 aluminum alloy.

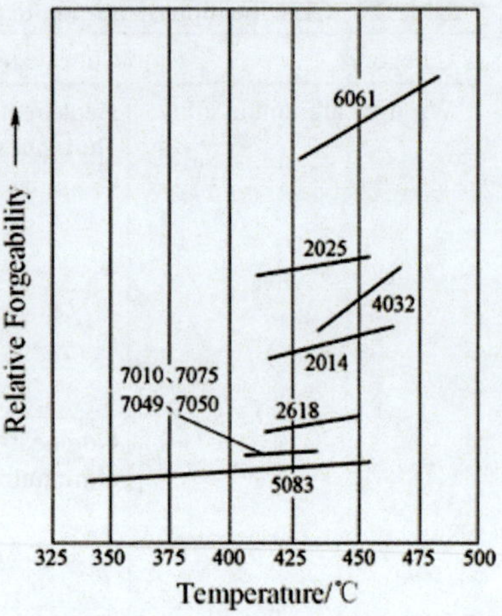

Fig. 2.1 Forgeability of ten aluminum alloys

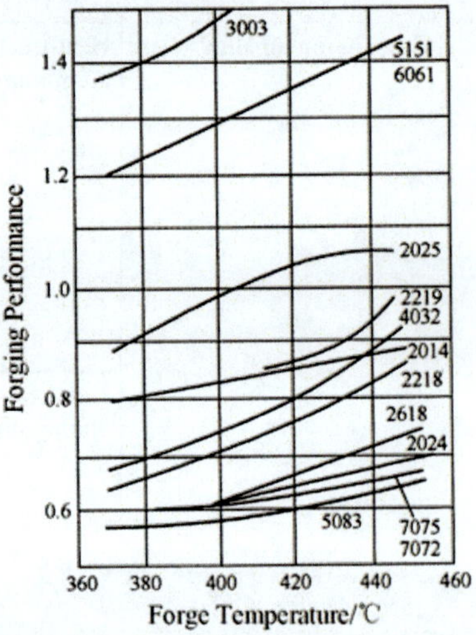

Fig. 2.2 Forging performance of some aluminum alloys

2.3 Aluminum Alloys and Their Mechanical Properties

Commonly used wrought aluminum alloys and their mechanical properties are listed in Table 2.2.

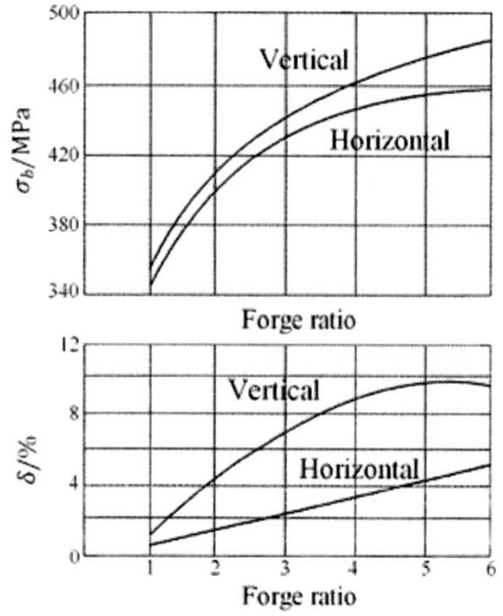

Fig. 2.3 The relationship between forging ratio and mechanical properties of 2014-T6 aluminum alloy

2.4 Forging Process Performance and Specifications

2.4.1 Forging Process Performance Analysis

The forging performance of some wrought aluminum alloys has been briefly elaborated above. Because different aluminum alloys have large differences in forging process performance, it is necessary to make a more detailed analysis.

The friction factor of aluminum alloys is generally considered to be three times that of steel, resulting in large plastic flow resistance. The differences of aluminum alloys in terms of forging process performance are as follows [1]:

(1) 6XXX series aluminum alloys

6XXX series aluminum alloys generally have medium strength and good plasticity. Take 6082 aluminum alloy that is used more in automobiles as an example, its tensile strength is greater than 310 MPa, yield strength is greater than 260 MPa, and elongation is greater than 10%. It is suitable for manufacturing automobile control arms and tie rods by multi-step die forging.

(2) 2XXX and 7XXX series aluminum alloys

Take the most used high-strength 7075 aluminum alloy as an example. When it is in the T6 state, its tensile strength is 600 MPa, the yield strength is 550 MPa, and the elongation is 12%. It has high strength, poor plasticity, large deformation resistance, and strong rate sensitivity. Since it is prone to cracks during free upsetting, drawing and open die forging, closed precision forging process should be employed.

(3) 4XXX series aluminum alloys

Take the medium silicon and medium–high silicon aluminum alloys as examples, because the Si content is 6.5–7.5% and 9–13% respectively, the strength in the solid

Table 2.2 Commonly used wrought aluminum alloys and their mechanical properties

Alloy grade	Material state	Elasticity modulus E/GPa	Shear modulus G/GPa	Poisson ratio μ	Tensile strength R_m/MPa	Specific plastic extension strength $R_p0.2$/MPa	Fatigue strength (the number of cycles is 5 × 10^8 times)/MPa	Shear resistance strength σ_τ/MPa	Elongation A_{11}/%	Section reduction ratio Z/%	Impact toughness a_k/(J/cm^2)	Hardness HBW
6A02	O	71	27	0.31	180	–	45	80	30	65	–	30
	T4	71	27	0.31	220	120	75	–	22	50	–	65
	T6	71	27	0.31	330	280	75	210	16	20	–	95
2A50	T6	71	27	0.31	420	300	–	–	13	–	–	105
2B50	Die forging T6	71	27	0.33	410	320	–	260	–	40	–	–
2A70	T6	71	27	0.31	440	330	–	–	12	–	–	120
2B70	T6	71	27	0.31	4740	270	–	–	10	–	–	120
2A90	T6	71	27	0.31	440	280	100	–	13	–	–	115
2A14	T6	72	27	0.33	490	380	115	290	12	25	10	135
2B12, 2A12	O	71	27	0.31	210	110	–	–	18	35	–	42
	T4	72	27	0.33	520	380	140	300	13	15	–	131
	Bar (φ40mm) T4	72	27	0.33	500	380		260	10	15	–	131
2A16	Extruded semi-finished products T6	71	27	0.31	400	250	130	–	13	35	–	110
	Sheet T6	71	27	0.31	420	300	.–	–	12	–	–	–

(continued)

Table 2.2 (continued)

Alloy grade	Material state	Elasticity modulus E/GPa	Shear modulus G/GPa	Poisson ratio μ	Tensile strength R_m/MPa	Specific plastic extension strength $R_p0.2$/MPa	Fatigue strength (the number of cycles is 5 × 10^8 times)/MPa	Shear resistance strength σ_τ/MPa	Elongation A_{11}/%	Section reduction ratio Z/%	Impact toughness a_k/(J/cm^2)	Hardness HBW
2A17	Forgings below 5 kg T6	71	–	–	430	350	–	–	9	18	–	–
2A01	O	71	27	0.31	16	60	–	–	24	–	–	38
	T4	71	27	0.31	300	170	95	200	24	50	–	70
2A02	Extrusion product T6	71	27	0.31	490	330	–	–	20	–	–	115
	Stamped impeller T6	71	27	0.31	440	300	–	–	15	–	–	115
2A04	Wire T4	70	–	–	460	280	–	290	23	42	–	115
2A06	Clad aluminum sheet T4	68	–	–	440	300	–	–	20	–	–	–
	Clad aluminum sheet HX4	68	–	–	540	440	–	–	10	–	–	–
2A10	T4	71	27	0.31	400	–	–	260	20	–	–	–
2B11, 2A11	O	71	27	0.31	210	110	75	–	18	58	30	45
	T4	71	27	0.31	420	240	105	270	15	30	–	100
7A03	Wire T6	71	–	–	520	440	–	320	15	45	–	150

(continued)

Table 2.2 (continued)

Alloy grade	Material state	Elasticity modulus E/GPa	Shear modulus G/GPa	Poisson ratio μ	Tensile strength R_m/MPa	Specific plastic extension strength $R_p0.2$/MPa	Fatigue strength (the number of cycles is 5 × 10^8 times)/MPa	Shear resistance strength σ_τ/MPa	Elongation A_{11}/%	Section reduction ratio Z/%	Impact toughness a_k/(J/cm^2)	Hardness HBW
7A04	T6	74	27	0.33	600	550	160	–	12	–	11	150
	O	74	27	0.33	260	130	–	–	13	–	–	–
	Clad aluminum sheet T6	74	27	0.33	540	470	–	–	10	23	–	–
	Clad aluminum sheet O	74	27	0.33	220	110	–	–	18	50	–	–
5A02	O	70	27	0.30	190	100	120	125	23	64	90	45
	HX4	70	27	0.30	250	210	130	150	6	–	–	60
5A03	O	70	27	0.30	200	100	110	155	22	–	–	50
	HX4	70	27	0.30	250	130	120	160	3	–	–	70
5A05	O	70	27	0.30	260	140	140	130	22	–	–	65
	HX4	70	27	0.30	300	200	–	–	14	–	–	80
	HX8	70	27	0.30	420	320	155	220	10	–	–	100
5A06	O	68	–	–	325	170	130	210	20	25	–	70
5A12	O	72	–	–	430	220	–	–	25	–	31	–
	F	–	–	–	580	500	–	–	10	–	–	–
5B05	O	70	27	0.30	270	150	–	190	23	–	–	70

(continued)

Table 2.2 (continued)

Alloy grade	Material state	Elasticity modulus E/GPa	Shear modulus G/GPa	Poisson ratio μ	Tensile strength R_m/MPa	Specific plastic extension strength $R_p0.2$/MPa	Fatigue strength (the number of cycles is 5×10^8 times)/MPa	Shear resistance strength σ_τ/MPa	Elongation A_{11}/%	Section reduction ratio Z/%	Impact toughness a_k/(J/cm^2)	Hardness HBW
3A21	O	71	27	0.33	130	50	55	80	23	70	–	30
	HX4	71	27	0.33	160	130	65	100	10	55	–	40
	HX8	71	24	0.33	220	180	70	110	5	50	–	55

state is relatively high, and the deformation resistance is still relatively high even when heated to the forging temperature. For this type of aluminum alloy, when the structure of the part is relatively simple, the extruded bar can be used for closed precision forging at high temperature; when the structure of the part is complex, such as an automobile steering knuckle, the combined casting-forging process can be employed [2].

(4) 5XXX series aluminum alloys

Because of its good plasticity, cold extrusion or cold precision forging can be employed to form the required parts, such as heat sinks.

2.4.2 Heating Specification

The forging temperature and heating specifications of different wrought aluminum alloys are listed in Table 2.3.

Table 2.3 Aluminum alloy forging temperature and heating specification

Alloy series	Alloy grade	Forging temperature/°C		Heating temperature/°C	Time required for heat penetration per unit thickness/(min/mm)
		Initial temperature	Final temperature		
6XXX series aluminum alloy	6A02	480	380	480	1.5
2XXX series aluminum alloy	2A50, 2B50, 2A70, 2A80, 2A90	470	360	470	
	2A14	460	360	460	
	2A01, 2A11, 2A16, 2A17	470	360	470	
	2A02, 2A12	460	360	460	
7XXX series aluminum alloy	7075, 7A09	450	380	450	3.0
5XXX series aluminum alloy	5A03	470	380	470	1.5
	5A02	470	360	470	
	5A06	470	400	400	

It can be seen from Table 2.3 that the forging temperature range of aluminum alloy is relatively narrow, generally within the range of 150 °C, and the forging temperature range of some high-strength aluminum alloys is even within the range of 100 °C. In addition, it should be pointed out that the forging temperature on the hammer is generally 20–30 °C lower than the forging temperature on the press.

2.4.3 *Allowable Deformation Degree and Forging Velocity*

The allowable deformation degree of different wrought aluminum alloys on different forging equipment is listed in Table 2.4.

Deformation velocity does not have much effect on the plasticity of most aluminum alloys, but the plasticity of a few highly alloyed aluminum alloys (such as 7075 aluminum alloy) decreases significantly when they are deformed at a high velocity. However, in order to increase the allowable degree of deformation and reduce deformation resistance and improve the fluidity of alloy filling die cavity, it is better to use hydraulic press for die forging than forging hammer, especially for large parts. In addition, due to the large friction factor and poor fluidity of aluminum alloy, if the deformation velocity is too fast, it is easy to generate defects such as skinning, folding and uneven grain size in the forged parts. For low-plasticity and high-strength aluminum alloys, it is also easy to cause cracking of the forging. Therefore, aluminum alloys are most suitable for forging on low-velocity presses.

2.5 Types of Aluminum Alloy Forgings

In order to facilitate the design of precision forging process and dies, the forged parts should be classified. The classification method of aluminum alloy forgings is similar to that of ferrous metal forgings, that is, in terms of the shape of the part. The

Table 2.4 Allowable deformation degree of commonly used aluminum alloy

Alloy	Hydraulic press	Forging hammer, hot die forging press	High speed hammer	Extrusion
	Upsetting			
Low strength and 2A50 alloy	80–85%	80–85%	80–90%, for 5A05 alloy 40–50%	90% and above
Medium strength	70%	50–60%	85–90%, for 5A06 alloy 40–50%	
High strength	70%	50–60%	85–90%	
Powder alloy	30–50%	50–60%	–	More than 80%

forgings with similar shapes basically employ the same precision forging process and die structure. At present, the general classification method is to divide aluminum alloy forgings into long shafts type, complex revolving type, branch-shaped type and large rib-web type according to the shape of the forgings and the axis direction of the billet during precision forging. More details are listed in Table 2.5.

2.5.1 *Long Shaft Forgings*

The characteristic of long shaft forgings is that the ratio of the length to the height or width is relatively large. According to the shape, axis line and parting characteristics of forgings, it can be divided into long rods, curved long rods and complex long rods. During die forging of the forgings, the axis of the billet is perpendicular to the striking direction of the forging force. The metal flows only in the height and width directions in the cross section (called the flow plane), and the flow along the axis is very small. This is due to the flow resistance of metal along the length direction is greater than that along lateral direction, which is a characteristic of plane strain. This feature provides a theoretical basis for the determination of the billet size of the long shaft forgings and the design of the billet-making and precision forging process. The flow model is shown in Fig. 2.4a.

2.5.2 *Complex Revolving Forgings*

It is also called short-shaft forgings, and its characteristic is that the height is generally smaller than the length and width in the top view, and the top view of the forgings is circular or approximately circular. When this type of forging is forged, the axis of the billet is consistent with the direction of the forging force, and the metal flows only in the longitudinal section (called the flow plane) along the height and the radial direction at the same time, which is the characteristic of axisymmetric deformation. This feature also provides a theoretical basis for the determination of the billet size of the revolving forging, and the design of the billet-making and precision forging process. The flow model is shown in Fig. 2.4b.

2.5.3 *Branch-Shaped Forgings*

For branch-shaped parts, the die forging process is also designed according to the feature of the long shaft forgings. It is worth noting that a pre-forging step for splitting material must be carried out after the billet-making step. When one branch is longer than the other branch and has a larger cross section, it is still regarded as a long shaft forging for die forging process design. If the length of all the branches is not much

Table 2.5 Classification of aluminum alloy forgings

Category	Group	Serial number	Part diagram
Long shaft type	Long rod	1–1	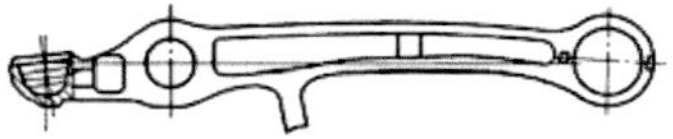
	Curved long rod	1–2	
	Complex long rod	1–3	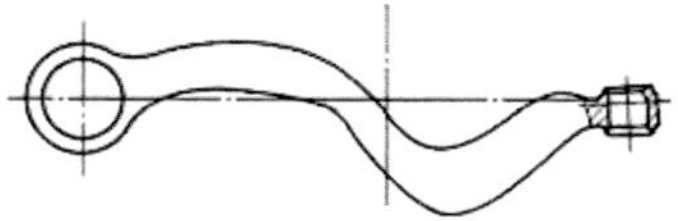
Complex revolving type	Solid	2–1	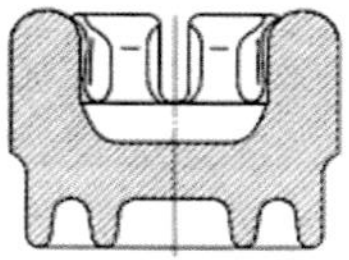
	Multilayer cylinder	2–2	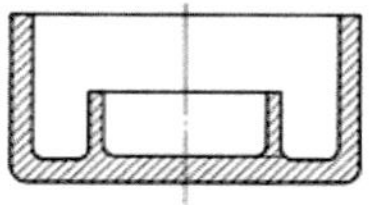

(continued)

Table 2.5 (continued)

Category	Group	Serial number	Part diagram
	Spiral cylinder	2–3	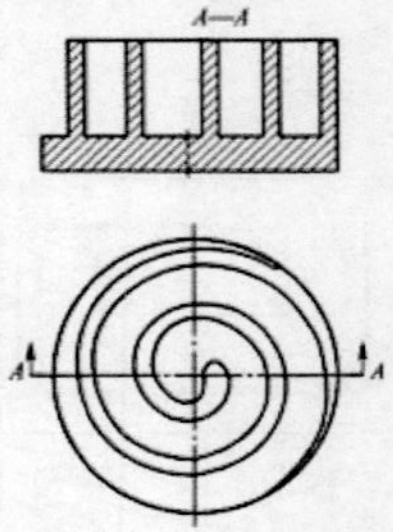
	Thin-walled cylinder	2–4	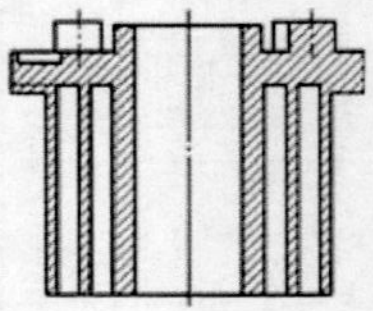
Branch-shaped type	Fork-shaped	3–1	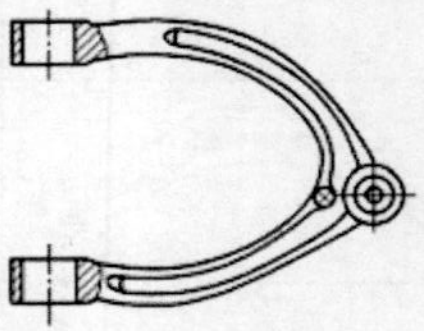

(continued)

Table 2.5 (continued)

Category	Group	Serial number	Part diagram
	Branch-shaped	3–2	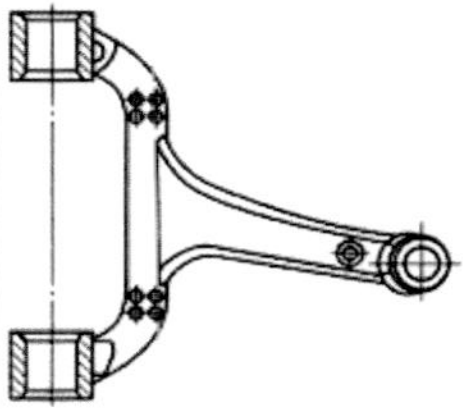
Large rib-web type	Long rib	4–1	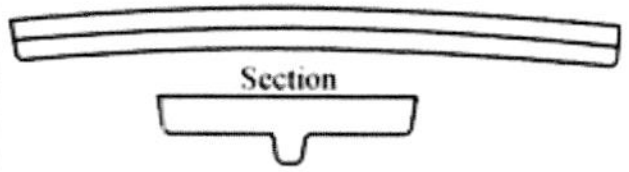
	Large projection	4–2	

Fig. 2.4 Metal flow model during forging: **a** long shaft forgings, **b** complex revolving forgings, **c** direction of flow

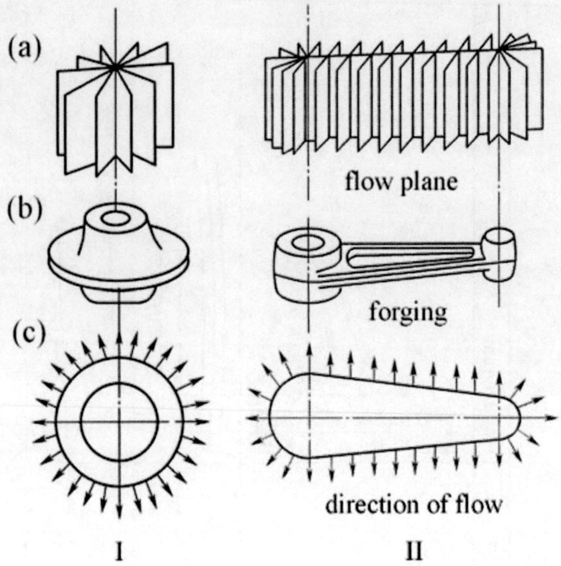

different, the semi-closed upsetting process can be used to prepare the billet, and the direct final forging process or first pre-forging and then final forging is carried out.

2.5.4 *Large Rib-Web Forgings*

Large rib-web forgings are mainly used in aircraft and aerospace vehicles. The feature of this type of forging is that ribs or cross ribs are vertically distributed on a long arc-shaped plate or a large projected flat plate. For these two kinds of rib-web parts, thick plates can be directly used for precision forging. Bars also can be used for forging after being flattened. Regarding whether the pre-forging step is required, it is determined according to the ratio of the height to the thickness of the ribs. The method of designing the process plan for precision forging is similar to that of long shaft forgings. When the ratio is large, a pre-forging step is required.

2.6 Forming Process of Aluminum Alloy Forgings

2.6.1 *Forming Process During Open Die Forging*

Open die forging is the most common final forging process, which is suitable for the final forging forming of various forgings with different structures. In order to compare with closed die forging, the revolving forging is selected as the research object to analyze its forming process and characteristics.

As shown in Fig. 2.5, the flow of deformed metal during open die forging is not completely restricted by the die cavity, and the excess metal will flow in the direction perpendicular to the force to form flash [3]. As the flash thickness and the forging temperature decreases, the forging force increases, and the outward flow of the metal is blocked by the flash, which eventually forces the metal to fill the die cavity.

In order to analyze the deformation process of open die forging, the whole deformation process can be divided into four stages.

Stage I (Fig. 2.5a, upsetting). The billet undergoes upsetting deformation in the die cavity, which may be accompanied by local press-in deformation for certain shapes of forgings. When the metal comes into contact with the side wall of the cavity, this stage ends. At this time, the deformed metal is in a weak three-dimensional compressive stress state, and the deformation resistance is also small.

Stage II (Fig. 2.5b, flash formation). In the later of the first stage, the metal flow is obstructed by the cavity wall, and the flow of the metal in the direction perpendicular to the forging force is restricted. When the compression continues, the metal flows to the depth of the cavity along the force direction, and meanwhile, continues to flow to the flash groove along the direction perpendicular to the force and form a little flash. At this time, the deformation resistance is obviously increased, and the metal in the cavity is in a strong three-way compressive stress state.

Stage III (Fig. 2.5c, the cavity is filled). After the flash is formed, the flash gradually becomes thinner as the deformation continues, and thus the resistance of the metal to the flash groove increases sharply. When the resistance of the flash groove is greater than the resistance of the metal flowing to the depth of the die cavity, the metal is forced to continue to flow to the depth of the die cavity until the entire cavity is completely filled. At this stage, the deformed metal is in a stronger three-dimensional compressive stress state, and the deformation resistance increases sharply.

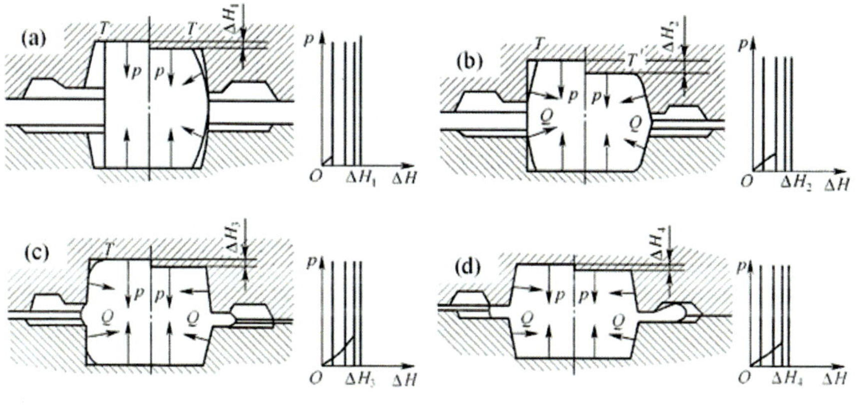

Fig. 2.5 The metal flow during open die forging: **a** upsetting, **b** flash formation, **c** the cavity is filled, **d** upper die and lower die contact

Stage IV (Fig. 2.5d, upper die and lower die contact). Generally, the volume of the billet is slightly larger than the volume of the die cavity. Therefore, when the die cavity is completely filled, the metal will continue to be compressed until the upper die and lower die are in contact with each other, that is, hitting. Excess metal is discharged into the flash groove to ensure that the height meets the requirements. The deformation at this stage only occurs in the area near the parting surface. At this stage, due to the further decreasing of the flash thickness, the resistance of the excess metal flowing out of the bridge of flash groove is very large. Thus, the deformation zone is in the strongest three-directional compressive stress state, and the deformation resistance is also the largest. Although the reduction in this stage is usually less than 2 mm, it consumes 30–50% of the total energy.

2.6.2 *Forming Process During Closed Die Forging*

Closed die forging is a forming process that the billet is deformed in a closed die cavity. The principle and process of closed die forging of the revolving part the in the integral die are as shown in Fig. 2.6. The forming process can be divided into three stages.

(1) Stage of upsetting (Fig. 2.6a) is from the moment of contact between the metal and the surface of the punch to the moment of contact between the metal and the side wall of the cavity. The upsetting during closed die forging is divided into two types: overall closed upsetting and partial closed upsetting. The former is positioned by the outer diameter of the billet, and the latter is positioned by the non-deformed part of the billet.

(2) Stage of filling the corner gap (Fig. 2.6b) is from the moment of contact of the drum-shaped side surface of the billet with the side wall of the die, until the entire side surface is attached to the die wall and the corner gap of the die cavity is completely filled. In this stage, the flow of the deformed metal is

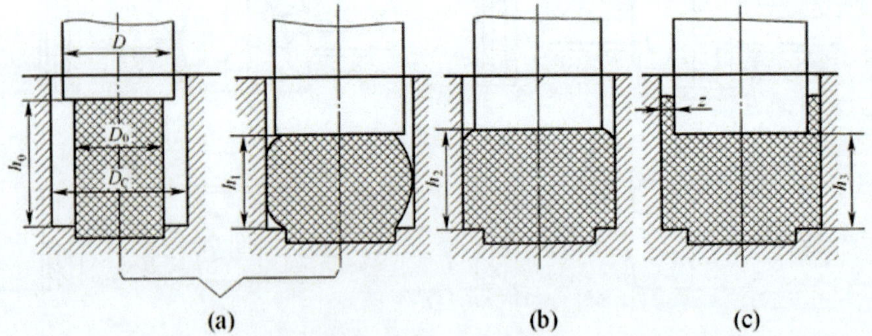

Fig. 2.6 The metal flow during closed die forging: **a** upsetting, **b** filling the corner gap, **c** squeezing into the flash cavity

hindered by the die wall, and each part of the deformed metal is in a different three-directional compressive stress state. As the degree of deformation of the billet increases, the lateral pressure on the die wall gradually increases until the die cavity is completely filled.

(3) Stage of squeezing into the flash cavity (Fig. 2.6c) is the excess metal is squeezed into the gap between the punch and the die under the action of increasing pressure to form longitudinal thin burrs.

Comparing Figs. 2.6 and 2.5, it can be seen that for the same revolving forging, lateral flash is generated around the forging during open die forging, while no lateral flash is generated around the forging during closed die forging, only longitudinal thin burrs are generated, which is beneficial to improve material utilization.

2.7 Calculation of Precision Forging Force

The purpose of the calculation of precision forging force is to select the tonnage of the precision forging equipment and provide a basis for the calculation of the strength and stiffness of the forging die.

2.7.1 Theoretical Calculation of Forging Force of Cylindrical Forgings

(1) When there is no flash at the end, assuming that the bottom gap of the cavity is finally filled, the deformation zone can be simplified to the sphere surrounded by a spherical surface with a radius of ρ and a thickness of h and an inclined free surface, as shown in Fig. 2.7. An element body (the shaded part in the figure) is taken from the deformation zone, the uniformly distributed stress acting on it is σ_r, σ_θ, $\sigma_r + d\sigma_\theta$ and τ_0, and the equilibrium differential equation of the element body is established in the direction of θ. Using the plastic condition and boundary conditions, the simplified expression of the unit pressure can be obtained:

$$p = \sigma_s[1 + \frac{2\alpha_1 R}{9a}(\frac{2R}{2R - a} - \frac{2a}{b})] \quad (2.1)$$

where σ_s is the yield strength of the metal listed in Table 2.6; α_1 is the angle between the free surface of the deformation zone and the die wall; R is the radius of the die cavity; a and b is the radial and height dimension of unfilled zone in the corner, respectively.

(2) When the longitudinal flash appears at the end, the deformation process is the same as that of backward extrusion, and the effect of flash needs to be

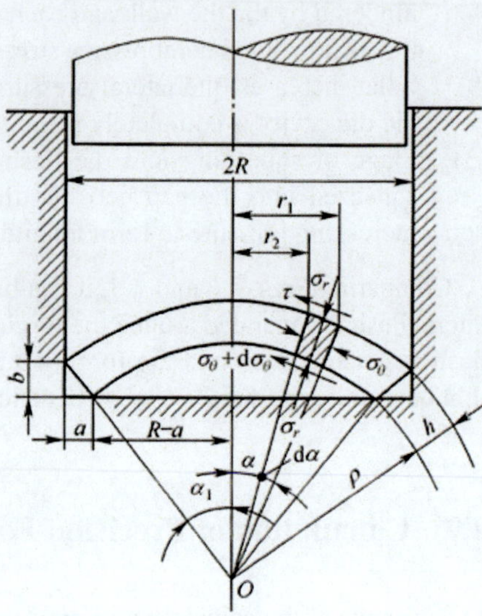

Fig. 2.7 Stress analysis when there is no flash at the end

Table 2.6 The yield strength of common used aluminum alloy. Unit: MPa

Alloy	200 °C	250 °C	300 °C	350 °C	400 °C	450 °C	500 °C
5A02, 5052			80	60	30	25	20
6A02, 6061	72	52	39	33	29	20	15
2A50, 2B50				57	40	32	25
2A70, 2B70, 2618			135	75	45	28	20
2A80			90	60	40	30	20
2A14, 2014			140	130	90	75	30
2A11, 2A12, 2024			110	75	55	40	25
7075, 7A09			90	70	55	40	35
3A21, 3003, 3004			40	30	25	20	15
2A02			210	120	80	50	20

considered when calculating the forging force. An element body is taken form the flash, as shown in Fig. 2.8, the normal stress in the z-direction and the x-direction can be obtained from the balance equation, plastic conditions and boundary conditions:

$$\sigma_z = \frac{4\mu\sigma_s}{D-d}(z-\lambda) \tag{2.2}$$

$$\sigma_x = \frac{4\mu\sigma_s}{D-d}(z-\lambda) - \sigma_z \tag{2.3}$$

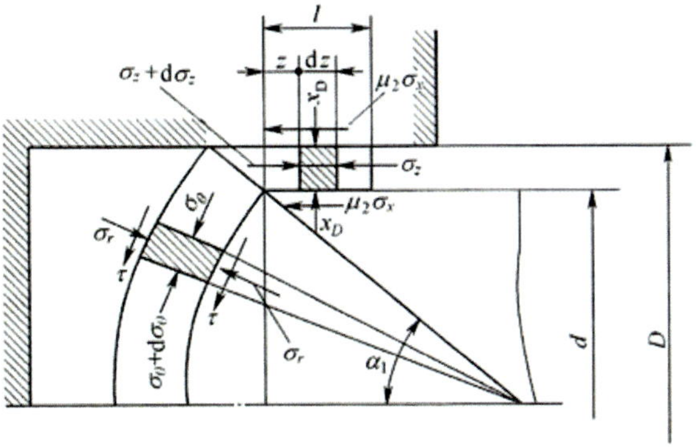

Fig. 2.8 Stress analysis when the longitudinal flash appears at the end

Then the simplified expression of the unit pressure can be deduced:

$$p = \sigma_{0.2}\left[1.7 + \frac{2.7\mu\lambda}{D - d} + \frac{\alpha_1 D}{4.5(D - d)}\right] \quad (2.4)$$

where μ is the friction coefficient on the contact surface between the deformed metal and the punch; λ is the height of the longitudinal flash; D is the diameter of the die cavity; d is the diameter of the punch; σ_s is the yield strength of the deformed metal, as listed in Table 2.6.

2.7.2 *Calculation of Forging Force During Small Flash Precision Forging*

The following equation is suitable for the calculation of the forging force of small flash precision forging on various forging equipment.

$$F = An_v\sigma_s n_d \quad (2.5)$$

where A is the contact area between the tool and the metal after deformation (including the contact area of small flash); n_v is the velocity coefficient, hydraulic press 1.0–1.1, crank press 1.0–1.3, screw press 1.3–1.5, forging hammer 2–3; σ_s is the yield strength of the deformed metal; n_d is the unit pressure coefficient that can be calculated by the following equations.

When round forgings is upset,

$$n_d = 1 + \frac{\mu d}{3h} \quad (2.6)$$

When rectangular forgings is upset,

Table 2.7 Friction coefficient of various materials

Material	The ratio of the deformation temperature to the melting temperature		
	0.8–0.95	0.5–0.8	0.3–0.5
Carbon steel	0.4–0.35	0.45–0.40	0.35–0.30
Aluminum alloy	0.5–0.48	0.48–0.45	0.35–0.30
Magnesium alloy	0.40–0.35	0.38–0.32	0.32–0.24
Heavy non-ferrous alloys	0.32–0.30	0.34–0.32	0.26–0.24
Non-ferrous heat-resistant alloy	0.28–0.25	0.26–0.22	0.24–0.20

Note The data in the table is for non-lubricated conditions. When lubricant is used, its value can be reduced by 25–15%

$$n_d = 1 + \frac{(3b - a)\mu a}{6bh} \tag{2.7}$$

When axisymmetric parts is forged,

$$n_d = \frac{\left(1 + \mu\frac{c}{h}\right)F_{mb} + \left(1 + 2\mu\frac{c}{h}\right)F_{dj}}{F_{mb} + F_{dj}} \tag{2.8}$$

where d is the diameter of the forging; h is the height of the forging or the height of the flash groove; a and b are the lengths of the two sides of the rectangular forging, and $a \leq b$; c is the width of the flash groove bridge; F_{mb} and F_{dj} are the horizontal projected area of flash groove bridge and die forging, respectively; μ is the friction coefficient (reference Table 2.7).

2.8 Deformation Mechanism of Aluminum Alloy

Aluminum alloys are polycrystalline like most other metal materials in which there are grain boundaries between grains, and sub-grains inside the grains. Therefore, in the precision forging, the hot deformation mechanism is more complicated than that of single crystal metal materials [4].

2.8.1 Basic Microstructure Concepts of Aluminum Alloy

(1) Lattice and face-centered cubic

In order to facilitate the understanding and description of the arrangement of atoms in the crystal, some straight lines are often used to connect the centers of the atoms in the crystal to form a spatial grid which is called a lattice. Usually a minimum geometric unit that can reflect the characteristics of the crystal lattice is selected to analyze the arrangement of atoms in the crystal. This minimum geometric unit is called a unit cell. The edge length of the unit cell is called the lattice constant, and the unit of measurement is usually 10^{-10} m.

Aluminum is a metal with a face-centered cubic structure. As shown in Fig. 2.9, there is an atom at each corner of the face-centered cubic unit cell, and an atom at the center of each face.

(2) Dislocation

The dislocation is a kind of linear crystallographic defect or irregularity within a crystal structure which contains an abrupt change in the arrangement of atoms. It can be generally divided into edge dislocation and screw dislocation. Among them, the schematic of edge dislocation is shown in Fig. 2.10. It can be seen that above a certain horizontal plane of the crystal, there is an extra vertical atomic plane, which is interrupted somewhere on the horizontal plane, like an inserted knife edge, causing an atomic dislocation between the above horizontal plane and the following two parts of the crystal. So it is called edge dislocation. The dislocation with more atomic planes in the upper half of the crystal is called the positive edge dislocation which is represented by the symbol "⊥". The dislocation with more atomic planes in the lower half of the crystal is called the negative edge dislocation which is represented

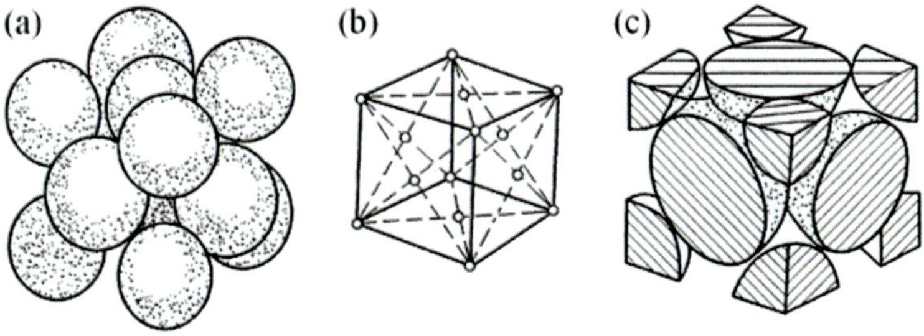

Fig. 2.9 Schematic of face-centered cubic structure: **a** rigid sphere model, **b** mass point model, **c** unit cell atomic number

Fig. 2.10 Schematic of edge dislocation

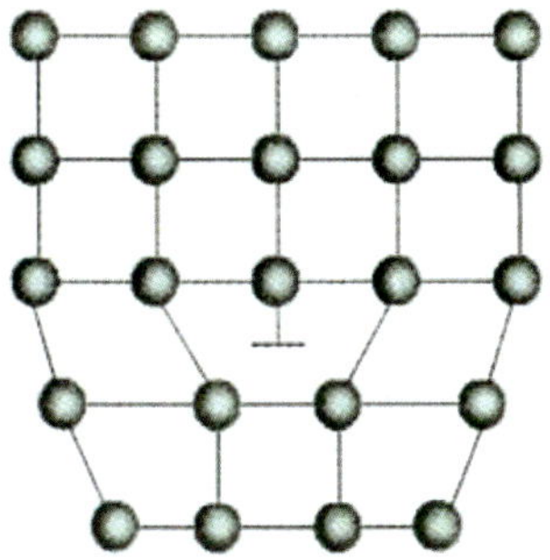

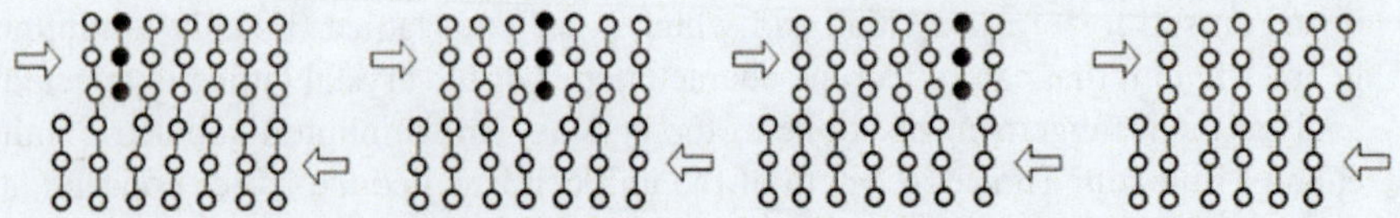

Fig. 2.11 Slip schematic of edge dislocation

by the symbol "⊤". It should be pointed out that the positive or negative of edge dislocation is relative.

(3) Slip

The plastic deformation of the crystal does not occur uniformly in the entire crystal. When the stress exceeds its elastic limit, the layers of the crystal generate relative displacement along a certain crystal plane and crystal direction, that is, slip [5], as shown in Fig. 2.11. This kind of displacement cannot be recovered after the stress is removed, and the accumulation of a large amount of slip constitutes a macroscopic plastic deformation. Slip is carried out under the action of shear stress. Whether a certain slip system of the crystal slips depends on the magnitude of the shear stress along the slip system. When a single crystal metal is stretched in the axial direction, the sliding plane that tends to the direction of 45° will slip first in a cubic structure metal with multiple sets of sliding plane.

(4) Twinning

The twinning occurs when a certain crystal plane in the crystal moves along a certain crystal direction. It is generally believed that twinning is a uniform shearing process that occurs inside the crystal. The displacement of each layer of crystal plane is proportional to the distance from the twin plane. The deformed part (twin) and the undeformed part (matrix) of the crystal is a mirror-symmetric relationship with the twin plane as the interface. Compared with slip, twinning is a sudden change process, the movement of the crystal is not necessarily an integer multiple of the atomic distance, and it is smaller than the movement of slip. In twinning, a part of the crystal undergoes uniform shear deformation, which is different from the slip concentrated in some plane planes. After twinning, the deformed part and undeformed part form a mirror-symmetrical position relationship, but after slip the relative position of each part of the crystal does not change. The twinning schematic of the face-centered cubic lattice structure is shown in Fig. 2.12.

2.8.2 Deformation Characteristics

When the single crystal is stressed, the external force can be decomposed into normal stress and shear stress on any crystal plane. Normal stress can cause elastic deformation and cleavage fracture of crystals, and shear stress causes plastic deformation

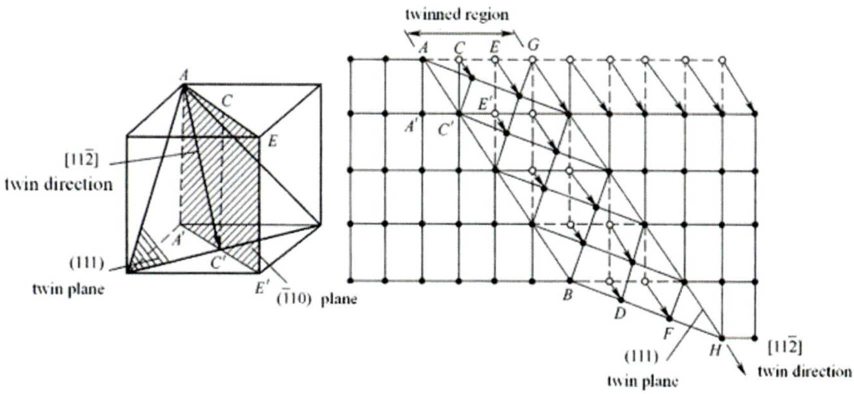

Fig. 2.12 Twinning schematic of face-centered cubic lattice

of metal crystals. For polycrystals, plastic deformation can only be produced under the action of shear stress.

Since polycrystals are composed of crystal grains with different crystal grain orientations, and there are grain boundaries between different crystal grains, the plastic deformation of polycrystals mainly includes intragranular deformation and intergranular deformation. The mechanism of intragranular deformation of polycrystals is consistent with that of single crystals. The main ways of intragranular deformation are slip and twinning. Slip refers to the sliding displacement of a part of the crystal relative to another part along a certain crystal plane and crystal direction. The twinning changes the orientation of the crystal lattice. The shear stress required for its deformation is much larger than the slip, and the deformation velocity is extremely fast, close to the speed of sound. The relative displacement between adjacent atoms during twinning is less than one atom distance.

The intergranular deformation is mainly the mutual sliding and rotation of the grains, as shown in Fig. 2.13. When the grains are deformed by an external force,

Fig. 2.13 Schematic of sliding and rotation between grains

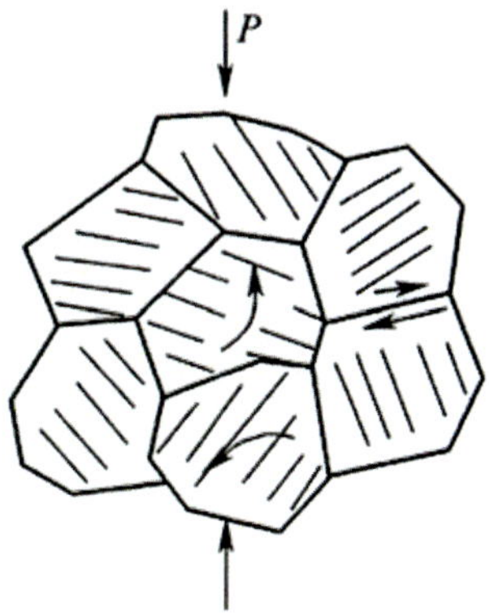

a shear stress may be generated along the grain boundary. When the shear stress is sufficient to overcome the relative sliding resistance between the grains, sliding occurs. In addition, when a force couple is generated between two adjacent grains, it will cause mutual rotation between the grains.

In polycrystals, the grains where the angle between slip system and external force is equal to or close to 45° occur slip first. When the stress at the front end of the plugging dislocation reaches a certain level, coupled with the rotation of the adjacent grains, the dislocations in the adjacent grains that were originally in the unfavorable position to the slip system move. Then the slip is transferred from a batch of grains to another batch of grains. When a large number of grains slip, the metal shows obvious plastic deformation.

In cold deformation, the plastic deformation of aluminum alloy is mainly intragranular deformation, and intergranular deformation only plays a minor role. When the grain boundary is deformed, it is easy to cause the destruction of the grain boundary structure and the generation of microcracks.

2.8.3 Mechanism of Hot Deformation

In hot deformation, the plastic deformation of aluminum alloy is mainly intragranular deformation and intergranular deformation. The main forms of intragranular deformation are slip and twinning, and the main forms of intergranular deformation are sliding and rotation between grains. In addition, during hot deformation, another deformation mechanism may appear, that is, diffusion. When the temperature rises, the thermal vibration of the atoms intensifies, and the atoms in the crystal lattice are in an unstable state. When the crystal is subjected to an external force, the atoms are continuously transferred from one equilibrium position to another in a non-synchronous manner along the direction of the stress field gradient (not along a certain crystal plane and crystal direction), causing plastic deformation of the metal. This kind of deformation is called diffusion plasticity. Diffusion plasticity requires a certain temperature and a certain time, and the general hot deformation velocity is relatively fast, and the diffusion plasticity is often too late to proceed.

During hot deformation, the strength of the grain boundary decreases, and the sliding and rotation between the crystal grains becomes very important. Grain boundary sliding proceeds along the direction with the greatest shear stress, and its shear process is uneven and discontinuous. After loading, the amount of sliding along different grain boundaries is different, even if the same place is at different times, the amount of sliding is different. The sliding of the grain boundary cannot be simply regarded as the relative sliding of the grains, but the result of deformation in a thin layer near the grain boundary. Because the deformation of aluminum alloy near the grain boundary is very large, it will soften first at high temperature, and the deformation can continue in these areas.

Hot deformation will have a great impact on the structure and properties of aluminum alloys. After the material is hot deformed, streamlines are formed due to inclusions, second phases, segregation, and grain boundaries distributed along the flow direction. The existence of streamlines will make the anisotropic mechanical properties of the part. Therefore, when designing the forging process, the streamline of part should be consistent with the direction of the maximum tensile stress to improve its load-bearing capacity.

The mechanical properties of the part after hot forging mainly depend on the grain size. The smaller the grain, the better the comprehensive mechanical properties. This requires precise control of hot forging process parameters including deformation temperature, deformation velocity and deformation degree.

2.8.4 *Microstructure Changes During Hot Deformation*

The temperature of hot deformation is generally above 0.6 times the melting point of the metal which is higher than the temperature of recrystallization. Therefore, recovery and recrystallization will occur during hot deformation. The recovery and recrystallization that occur during the deformation process are usually called dynamic recovery and dynamic recrystallization. It is generally believed that dynamic recovery and dynamic recrystallization are important mechanisms of metal softening during hot deformation. Figure 2.14 shows schematic of dynamic recrystallization and static recrystallization.

Dynamic recovery often occurs in the hot deformation process of some metals with high stacking fault energy, such as aluminum and aluminum alloys, ferritic steel, and zinc, magnesium, and tin. When this kind of metal is hot deformed, its dislocation slipping and climbing are relatively easy to proceed. Therefore, it is generally believed that dynamic recovery is the main softening mechanism during the hot processing of this type of material, even if it is much higher than the static recrystallization temperature. If it is rapidly cooled to room temperature after hot deformation, it can be found from the microstructure of the material that the grains become elongated in the deformation direction and become fibrous, and meanwhile, equiaxed sub-grains formed by dynamic recovery appear in the grains. The sub-grains are repeatedly disassembled and composed during the deformation process, and their size is controlled by the deformation temperature and the deformation velocity. With the decrease of deformation velocity or the increase of temperature, the sub-grain size increases. The dislocation density after dynamic recovery is higher than that of static recovery.

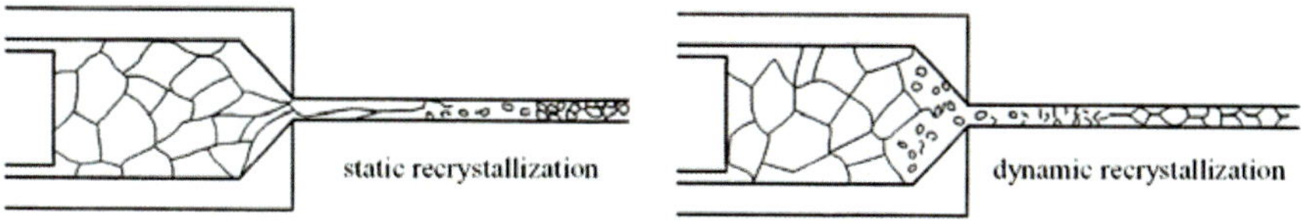

Fig. 2.14 Schematic of dynamic recrystallization and static recrystallization

The microstructure of dynamic recovery is much stronger than the recrystallized microstructure. Maintaining the microstructure of dynamic recovery after hot deformation has been successfully used to improve the strength of aluminum alloy extruded profiles. In addition, some aluminum alloys can obtain higher strength after heat treatment.

2.9 Quality Control of Aluminum Alloy Forgings

The quality of aluminum alloy forgings can be controlled from three stages: before forging, during forging and after forging.

2.9.1 Quality Control Before Forging

Quality control before forging is mainly to control the quality of raw materials. The main control contents are as follows:

(1) The chemical composition of raw materials should meet the requirements of the standard, and there should be no internal defects and compound segregation.
(2) The extruded bar or profile generally does not allow the as-cast microstructure inside.
(3) Both ends of the bar or profile are not allowed to have tail recesses or interlayers.
(4) The surface of the bar or profile should not have defects such as excessive knocks, scars, scratches and bruises.
(5) If aluminum alloy ingots are used for the production of large aluminum alloy forgings, upsetting and stretching should be used for billet-making.
(6) The serial number of heating and batch must be marked on the raw materials so that the quality of the forging can be traced.

2.9.2 Quality Control During Forging

Quality control during forging is the control of the forging production process which mainly includes the die forging process, process parameters, dies, equipment, lubrication and operation. The key points of quality control are as follows:

(1) For the billets used in the production of precision forgings, the size deviation or volume deviation, the flatness of the end surface and the perpendicularity of the end surface to the axis should meet the requirements of the precision forging.
(2) The billet should be heated uniformly, and the heating temperature and initial forging temperature should be strictly controlled.

(3) The range of forging temperature T is, $T_{start} \geq T \geq T_{end}$.
(4) The forging temperature range of aluminum alloy is narrow, generally within 100 °C. Therefore, the die should be preheated before forging. In order to ensure uniform preheating, resistance rods can used to inserted into the preheating holes on the dies.
(5) It is recommended to use small flash and no flash precision forging, especially for the aluminum alloy with high strength to avoid cracks.
(6) The three main process parameters (deformation degree, deformation velocity and temperature) should be optimized and the relationship between the three should be considered. For cold precision forgings, their performance after recrystallization annealing is basically the same as before cold deformation. Whether the performance of the forged part is exactly the same as before deformation depends on whether the microstructure before cold deformation and after recrystallization are the same. In order to control the performance of the part after deformation, the grain size of the metal after recrystallization is often controlled during production. There are many factors that affect the size of the recrystallized grains, mainly including the degree of deformation, the recrystallization temperature, the size of the original grains, and second phase particles. The relationship between the grain size after recrystallization annealing and the degree of cold deformation and recrystallization temperature is usually drawn as a solid figure, called a recrystallization diagram [6], as shown in Fig. 2.15. This figure has important reference value for controlling the grain size of aluminum alloy forgings during annealing after cold deformation.
(7) In cold precision forging and hot precision forging, the die must be lubricated.
(8) The production cycle should be uniform. The consistency of the operating cycle can ensure the stability of the process parameters and the consistency of the quality of the forgings.

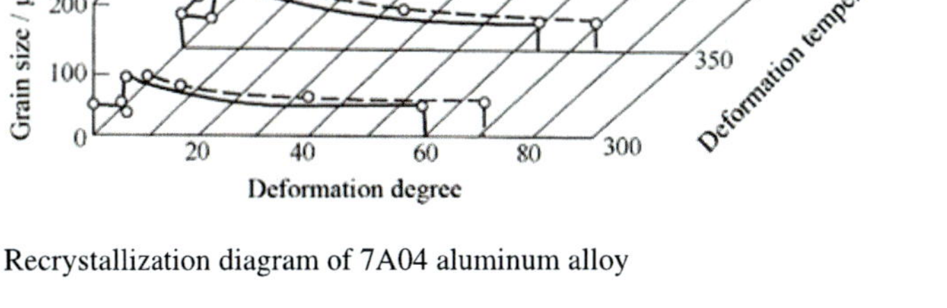

Fig. 2.15 Recrystallization diagram of 7A04 aluminum alloy

2.9.3 Quality Control After Forging

The main methods of quality control after forging are as follows:

(1) The residual layer of lubricant on the surface of the forging should be cleaned. The graphite emulsion lubrication is usually used in production of aluminum alloy forgings. Due to the high unit pressure during forging, the black graphite emulsion particles adhere firmly to the surface of the forging, which seriously affects the surface quality of the forging. It can be cleaned up with alkali solution, nitric acid solution and clean water.

(2) The surface scars should be repaired. Especially during hot forging, because aluminum alloys are very soft, the forgings are prone to generate surface defects such as partial tearing which should be scraped or polished to a smooth surface.

(3) The heat treatment should be carried out in strict accordance with the heat treatment specifications to meet the requirements of mechanical properties of forgings.

References

1. Wu SX, Pan QJ (2014) Handbook of wrought aluminum alloy and its die forging technology. China Machine Press, Beijing
2. Kim M, Lim T, Yoon K, Ko Y, Kim KJM, Kwak K (2011) Development of cast-forged knuckle using high strength aluminum alloy. SAE technical paper, 2011-01-0537
3. Xia JC, Zhang QX (2010) Material forming technology. China Machine Press, Beijing
4. Pan J, Tian M, Tong J (2011) Fundamentals of materials science. Tsinghua University Press, Beijing
5. Liu JA, Zhang HW, Xie SS (2012) Forging technology of aluminum alloy. Metallurgical Industry Press, Beijing
6. Yu HQ, Chen JD (1999) Principles of metal plastic forming. China Machine Press, Beijing

Chapter 3
Finite Element Simulation of Precision Forging

3.1 Overview

Physical experiment method and theoretical analysis method are traditional methods to study the law of metal plastic deformation, but there are obvious problems. The physical experiment method mainly refers to the trial and error method which is reliable, but the cost is too high and the cycle is long. The theoretical analysis method is only suitable for the forming analysis of parts with simple shape, and requires many assumptions, resulting in low accuracy and limited application. Under the situation of faster and faster production pace, and more complex shapes of parts, traditional research methods can no longer meet the requirements. With the advancement of computers and plastic deformation theory, numerical simulation has shown great advantages in studying metal plastic deformation. Numerical simulation can simulate the stress, strain, microstructure of metal plastic deformation by establishing analysis models, replacing the physical experiment and theoretical analysis method to study the law of plastic deformation.

The finite element method is a mature and widely used numerical simulation method, which originated in the early 1940s. In 1960, Clough formally used this term when analyzing the structural problems of two-dimensional plane stress. The basic idea of the finite element method is to discretize the continuous solution area with infinite degrees of freedom into discrete bodies (elements) with finite degrees of freedom and connected to each other in a certain way (nodes), that is, to divide the continuum hypothesis into a limited number of discrete units, and the units are only connected to each other at a limited number of designated points. The original continuum is replaced by a collection of discrete units. In general, finite element equations are a set of linear equations with node displacements as unknowns. Solving the equations can get the displacements at a finite number of nodes on the continuum, and then the distribution law of physical quantities such as stress on each element can be obtained.

L. Deng et al., *Precision Forging Technology and Equipment for Aluminum Alloy*, Springer Series in Advanced Manufacturing,
https://doi.org/10.1007/978-981-19-1828-5_3

In 1967, Marcal and King applied the elastic–plastic finite element method to plastic forming. In 1968, Yamada derived the stress–strain matrix of small elastoplastic deformation. This elastoplastic finite element method based on the theory of small deformation is not suitable for large plastic deformation, but it promotes the development of elastoplastic finite element for large deformation. The research on the basic theory of large deformation can be traced back to Hill's work in 1959, but it was not until 1970 that Hibbit et al. proposed elastoplastic finite element formulations of large deformation based on the principle of virtual work and the Lagrangian description. In the 1970s, Osias and McMeeking used Euler's description method to establish finite element formulations of large deformation. Since then, the elastoplastic finite element method of large deformation has been continuously developed and improved.

The elastoplastic finite element method is based on the theoretical basis of finite deformation, and uses an incremental constitutive relationship. The incremental step is very small, so the total calculation is very long. In order to overcome the problem of the elastoplastic finite element method, Lung introduced the volume incompressibility condition through the Lagrangian multiplier into the Markov variational principle to establish the rigid-plastic finite element formulation. Lee and Kobayashi respectively proposed a similar rigid-plastic finite element method by matrix analysis in 1973. Zienkiewicz used the penalty function method to introduce the volume incompressibility condition into the Markov variational principle in 1979, and obtained the corresponding rigid-plastic finite element formulation. Rigid-plastic finite element method is only suitable for cold deformation. In hot deformation, the strain hardening effect is weakened, and the deformation rate sensitivity increases, so the viscoplastic constitutive relationship needs to be used. Zienkiewicz regarded the metal in hot deformation as an incompressible non-Newtonian viscous fluid in 1972, and derived the rigid viscoplastic finite element formulation. Kobayashi and Oh also derived a similar finite element formulation on the basis of the variational principle of rigid viscoplastic materials.

Subsequently, the finite element simulation technology of metal plastic forming has been developed rapidly. The simulation object has developed from a simple two-dimensional model to a complex three-dimensional model, and from a macroscopic simulation to a microstructure simulation [1, 2]. After decades of development, a variety of finite element simulation software has emerged in the world, such as DEFORM, MSC.Marc, SIMUFACT, ABAQUS, etc.

3.2 Basic Theory of Finite Element Simulation

3.2.1 Rigid Viscoplastic Finite Element Method

The basic assumptions of the rigid viscoplastic finite element method for materials are as follows:

(1) The elastic deformation of the material and the influence of the volume force (gravity and inertial force, etc.) are not considered.
(2) The material is homogeneous and isotropic, and the volume is not compressible.
(3) The deformation of the material obeys the Levy–Mises flow theory.

Rigid viscoplastic materials should satisfy the following plastic equations and boundary conditions during plastic deformation.

(1) Balanced differential equation

$$\sigma_{ij,j} = 0 \tag{3.1}$$

where $\sigma_{ij,j}$ is the Cauchy stress tensor.

(2) Geometric equation

$$\dot{\varepsilon}_{ij} = \frac{1}{2} \cdot \left(\dot{u}_{i,j} + \dot{u}_{j,i} \right) \tag{3.2}$$

where $\dot{\varepsilon}_{ij}$ is the strain rate tensor; $\dot{u}_{i,j}$ is the velocity component.

(3) Constitutive relationship

$$\sigma'_{ij} = \frac{2}{3}\frac{\overline{\sigma}}{\dot{\overline{\varepsilon}}} \dot{\varepsilon}_{ij} \tag{3.3}$$

where σ'_{ij} is the stress deflection tensor; $\overline{\sigma}$ the equivalent stress, for rigid viscoplastic material, $\overline{\sigma} = \overline{\sigma}(\overline{\varepsilon}, \dot{\overline{\varepsilon}}) = \sqrt{\frac{3}{2}\sigma'_{ij}\sigma'_{ij}}$; $\dot{\overline{\varepsilon}}$ is the equivalent strain rate, $\dot{\overline{\varepsilon}} = \sqrt{\frac{2}{3}\dot{\varepsilon}_{ij}\dot{\varepsilon}_{ij}}$; $\dot{\varepsilon}_{ij}$ is the viscoplastic strain rate.

(4) Mises yield criteria

$$f = \overline{\sigma} - \sigma_s \tag{3.4}$$

where σ_s is the yield strength of the material, $\sigma_s = f\left(\dot{\overline{\varepsilon}}, \overline{\varepsilon}, T\right)$.

(5) Incompressible volume

$$\dot{\varepsilon}_{v} = \dot{\varepsilon}_{ij} \cdot \delta_{ij} = 0 \tag{3.5}$$

where $\dot{\varepsilon}_{v}$ is the volumetric strain rate; δ_{ij} is the Kronecker unit tensor.

(6) Boundary conditions, including stress boundary and velocity boundary conditions

$$\sigma_{ij} n_j = \overline{p}_i \tag{3.6}$$

$$\dot{\underset{i}{u}} = \dot{\underset{i}{u}} \tag{3.7}$$

where n_j is the component of the normal vector outside the unit at any point on the force surface S_p; $\overline{p}_i$ is the surface stress.

3.2.2 Thermal–Mechanical Coupling Finite Element Method

In hot plastic deformation of metal, there is heat exchange between the billet, dies and the environment. Meanwhile, dut to the change of plastic deformation work and friction, the geometric configuration of the temperature field, internal heat source, and temperature boundary conditions also change, which lead to the temperature field in the billet and the die is constantly changing. Therefore, the deformation analysis and thermal analysis must be coupled to simulate the forming process of metal reasonably and effectively [3].

The temperature field of plastic deformation is an unstable heat transfer problem with internal heat source. From the law of conservation of energy, the thermal analysis control equation is as follows.

$$\dot{U} = Q_c + Q_v + Q_\tau \tag{3.8}$$

where $\dot{U}$ is the change in internal energy of the workpiece; Q_c is the thermal energy converted by heat conduction; Q_v is the thermal energy converted by plastic strain energy; Q_τ is the thermal energy converted by the friction work of the die-workpiece interface.

After transforming Eq. (3.8), the following equation can be obtained,

$$\int_V C\rho \dot{T} dV = \int_V \lambda T_{,ii} dV + \int_V \alpha \overline{\sigma}\, \dot{\overline{\varepsilon}} dV + \int_S \beta |\tau_f||v_r| dS \tag{3.9}$$

where C is the specific heat capacity of constant volume; ρ is the material density; $\dot{T}$ is the rate of change of temperature versus time, $\dot{T} = \frac{\partial T}{\partial t}$; λ is the thermal conductivity of the workpiece; $T_{,ii}$ is the temperature gradient; α is the percentage of strain energy converted to heat energy, generally taken $\alpha = 0.9 - 0.95$; $\overline{\sigma}$ is the equivalent stress; $\dot{\overline{\varepsilon}}$ is effective strain rate; β is the heat distribution coefficient, generally taken $\beta = 0.5$; τ_f is the friction stress between the die and the workpiece; v_r is the relative sliding velocity between the workpiece and the die.

The coupling of deformation and heat is realized by the constitutive relationship of materials. For the plastic deformation process affected by temperature, deformation analysis and heat transfer analysis affect each other. The temperature change during the deformation process will cause the yield strength and some temperature-related material properties to change, and the change of the material properties will affect the analysis of the deformation process. The deformation process of the material greatly affects the temperature distribution, and also affects the thermal boundary conditions such as heat conduction, convection and radiation. Therefore, it is necessary to solve the plastic deformation velocity equation and the heat conduction equation under a given temperature distribution simultaneously, that is, to perform a coupled calculation of deformation and thermal analysis.

The calculation steps of thermal–mechanical coupling are as follows:

(1) Assume or calculate the initial temperature field.
(2) Calculate the initial deformation field according to the initial deformation conditions.
(3) Calculate the initial temperature change rate from the results of (1) and (2).
(4) Refresh node coordinates and element equivalent strain and related variables.
(5) Calculate the first-level approximate value of the temperature field based on the velocity field of the previous step.
(6) Calculate the new velocity field corresponding to the approximate value of the first-level temperature field.
(7) Use the new velocity field to calculate the temperature field and the second-level approximation.
(8) Repeat (6) and (7) until a convergent solution is obtained.
(9) Calculate the new temperature change rate.
10) Repeat (3)–(9) until the required deformation is reached.

3.3 Key Technologies of Finite Element Simulation

3.3.1 Calculation of Contact Friction

Assume that the friction between the workpiece and the die obeys Coulomb's law. Since the form of Coulomb's law is similar to a step function, when the relative velocity is very small and its direction changes during calculation, or when the contact point undergoes a conversion between the adhesion state and the sliding state, it is easy to cause the instability of the numerical calculation. Therefore, in order to avoid the sudden change of friction caused by the change of contact state, the modified Coulomb friction law is employed to make the friction continuously change with the relative sliding velocity [4].

By introducing the smooth function to modify the Coulomb friction law, the above calculation instability can be overcome. The hyperbolic tangent function with good asymptotic properties can be selected as the smoothing function, so the friction force can be calculated as follows

$$\vec{f}_\tau = -\mu P_n \tanh\left(\frac{\left\|\vec{u}_\tau\right\|}{d}\right)\frac{\vec{u}_\tau}{\left\|\vec{u}_\tau\right\|} \tag{3.10}$$

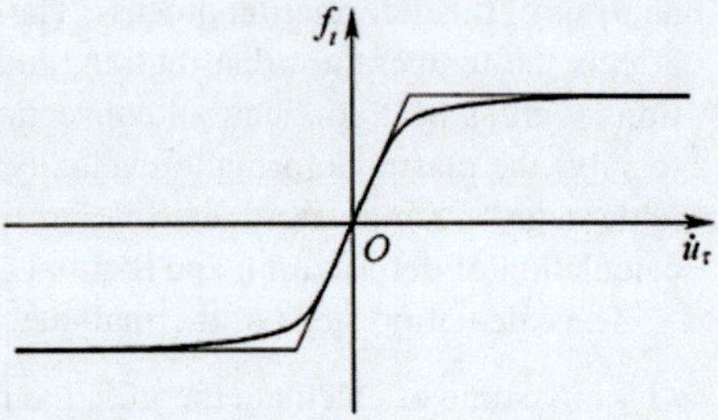

Fig. 3.1 Modified Coulomb friction law

where μ is the friction coefficient; P_n is the normal contact force; $\vec{\dot{u}}_\tau$ is the relative sliding velocity of the contact point between the workpiece and the die; d is a small positive number; $\left\| \vec{\dot{u}}_\tau \right\| = \sqrt{\dot{u}_{\tau 1}^2 + \dot{u}_{\tau 2}^2 + \dot{u}_{\tau 3}^2}$.

The relationship between the friction force and the relative velocity expressed by the Eq. (3.10) can be expressed by the continuous curve in the modified Coulomb friction law as shown in Fig. 3.1. Obviously, the relationship between friction and relative velocity becomes a smooth continuous function which effectively avoids the sudden change of friction caused by the change of contact state.

3.3.2 *Meshing Technology*

(1) Selection of element type

The solver of finite element software can usually support tetrahedral and hexahedral mesh elements. Generally speaking, hexahedral mesh has higher efficiency and calculation accuracy than tetrahedron. But for geometric bodies with complex shapes, it is quite difficult to divide them with hexahedrons. Especially in the forging process, flashes and burrs are often formed. At this time, it is undoubtedly unfeasible to use a hexahedron to divide the mesh. In contrast, tetrahedral elements have great advantages in this aspect. Tetrahedrons can be used to discretize very complex geometrical bodies, especially for curved surfaces, small features and irregular curved surfaces. In addition, it can be easily re-meshed and adaptively divided. Although the accuracy of the tetrahedral element is lower than that of the hexahedral element of the same size, the accuracy can be improved by appropriately increasing the number of elements.

(2) Adaptive meshing technology

When the mesh is divided, the density of the mesh in different regions can be adjusted by the weight factor, so as to obtain a relatively high-precision result. The physical variables that can be set to weight factors include curvature, temperature, strain, and strain rate. During the calculation process, a relatively fine mesh should be generated in the region with large surface curvature, temperature gradient, strain gradient and strain rate. In actual numerical simulation, the corresponding weighting factor is set according to the analyzed problem.

(3) Local mesh refinement technology

For small features of forgings, finer elements are needed to obtain accurate calculation results. However, if fine elements are divided in the overall forging, it will cause a large amount of calculation and corresponding low efficiency. To solve this problem, local mesh refinement can be performed. According to the needs of the analyzed problem, the refinement area is selected, and the size of the fine element is specified.

(4) Automatic re-meshing technology

In the process of numerical simulation calculation, the element deformation may be very large, which can easily lead to deformity. Excessive deformation will cause difficulty in the convergence of the calculation process. At this time, the element must be remeshed. By setting the trigger parameters for re-meshing, when any trigger parameter reaches the user-specified value, the solver will call the mesher to re-mesh. Thus a better mesh will be generated, and the value of variables on the original mesh will change. The value is interpolated onto the new element.

3.3.3 Choice of Solver and Iterative Algorithm

There are manily sparse matrix solver and conjugate gradient solver for finite element numerical simulation. The sparse matrix solver uses a direct solution method, and the conjugate gradient solver uses an iterative solution method. Conjugate gradient solver has great advantages over sparse matrix solver. The calculation speed is faster. For large-scale problems, the calculation time can be shortened by 4/5. The hardware requirements are lower. The large numerical calculations can be realized on ordinary computers. Larger-scale computing tasks can be handled. However, the conjugate gradient solver also has shortcomings. Sometimes, the sparse matrix solver may be used to converge smoothly, but the conjugate gradient solver is difficult to converge or even does not converge, especially when dealing with large rigid slip problems.

The advantages and disadvantages of the two solvers can be used comprehensively. When the workpiece is in contact with the die, because it is easy to produce large rigid body slip, the sparse matrix solver will be better. When a stable boundary is established, the conjugate gradient solver can be used to continue the calculation.

3.4 Examples of Material Modeling for Finite Element Simulation

In the application of finite element method, thermal–mechanical coupling finite element simulation is the most widely used for plastic forming. In addition, in recent years, the simulation of microstructure evolution has also attracted extensive attention. The following takes 2024 high-strength aluminum alloy as an example to focus on the establishment of high-temperature constitutive equations and dynamic recrystallization models [5].

3.4.1 *Constitutive Modelling of 2024 Aluminum Alloy*

(1) Hot compression experiment

Before establishing the high-temperature constitutive equation of the material, a high-temperature uniaxial compression experiment is required to obtain the stress value under different temperatures, strain rates, and strains. The material used in the experiment was 2024 aluminum alloy produced by the Southwest Aluminum (Group) Co., Ltd. in China. In order to make the microstructure of the material uniform and eliminate the residual stress, a complete annealing treatment was carried out. First, the resistance furnace was heated to 410 °C, and then the material was placed in the furnace for 2 h. Subsequently, the temperature was reduced to 270 °C with a rate of 30 °C/h. Finally, the material was taken out for air cooling. The microstructure of the annealed material showed long-axis crystal grains with an average grain size of about 25 μm.

The compression experiment was carried out on a Zwick/Roell Z020 universal material testing machine. The heating of the sample was realized by the heating furnace and temperature control system of the testing machine. The sample was heated to the deformation temperature at a heating rate of 10 °C/s, and kept for 3 min. Then compression deformation was carried out immediately. After compression, the sample was quickly taken out and water quenched to retain the high-temperature deformed microstructure. Temperature (T) and strain rate ($\dot{\varepsilon}$) are the key factors that affect the flow behavior of the material. The experimental temperature range was 300—450 °C, and the strain rate range was 0.001 s^{-1}–1 s^{-1}. The displacement and load of deformation were collected by the testing machine in real time, and the data can be directly converted into a true stress–strain curve.

(2) Stress–strain curve

The experimentally obtained 2024 aluminum alloy stress–strain curve is shown in Fig. 3.2. According to the trend, the curve can be divided into two situations. One is with obvious stress peak point, that is, after the stress reaches the peak, it gradually decreases due to the dynamic softening effect and reaches a stable state. This situation mainly occurs under low temperature or low strain rate, such as at 300 °C or 0.001 s^{-1}. The other is no obvious stress peak point, that is, the stress keeps increasing and finally reaches a stable state as the strain increases. This situation mainly occurs under high temperature or high strain rate.

It can also be seen from Fig. 3.2 that as the temperature increases or the strain rate decreases, the stress increases, and the stress peak appears earlier.

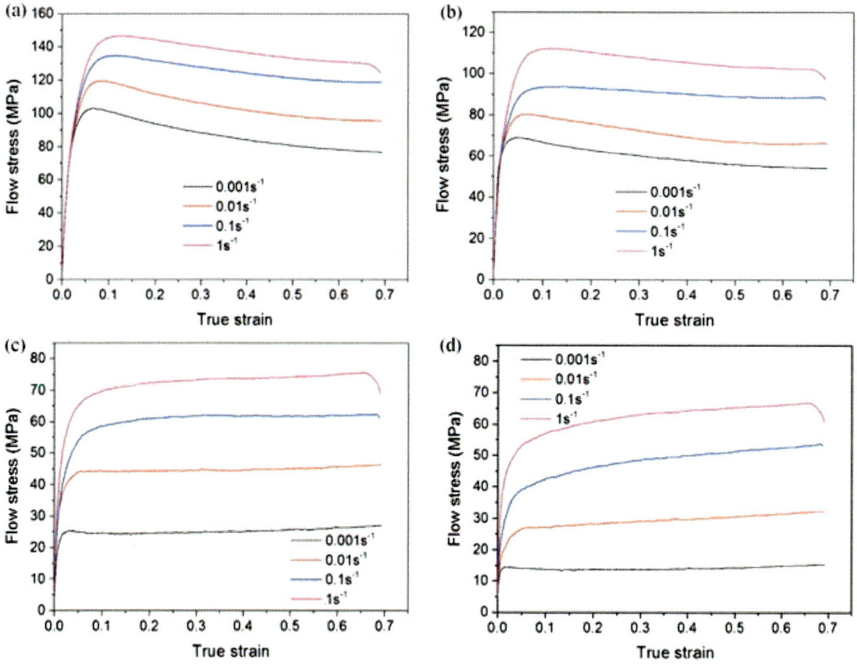

Fig. 3.2 Stress–strain curve: **a** $T = 300$ °C, **b** $T = 350$ °C, **c** $T = 400$ °C, **d** $T = 450$ °C

(3) Establishment of constitutive equation

Generally speaking, the flow stress of a material is mainly affected by its own characteristics (such as initial grain size, second phase content, etc.) and deformation conditions (such as deformation temperature, strain rate, deformation degree and deformation mode, etc.). Therefore, the constitutive equation of the material can be expressed as [6]:

$$\sigma = f(\varepsilon, \dot{\varepsilon}, T, C, S) \tag{3.11}$$

where σ is the flow stress; ε is the true strain; $\dot{\varepsilon}$ is the strain rate; T is the deformation temperature; C and S are the parameters related to the internal chemical composition and microstructure of the material, and the evolution of the microstructure is related to the deformation conditions.

When the material is determined, the flow stress can be expressed as a function of strain, deformation temperature and strain rate. So the Eq. (3.11) can be simplified as:

$$\sigma = f(\varepsilon, \dot{\varepsilon}, T) \tag{3.12}$$

In 1966, Sellars et al. proposed a hyperbolic-sine type model based on Eq. (3.12) [7], namely

$$\dot{\varepsilon} = A_1 F(\sigma)\exp(-Q/RT) \tag{3.13}$$

where $F(\sigma)$ is the function of stress; Q is the deformation activation energy; R is the gas constant.

There are different expressions in the constitutive equation for different $\alpha\sigma$ ranges. The Eq. (3.13) can be transformed into [8, 9]:

$$\begin{aligned} \dot{\varepsilon} &= A_1\sigma^{n_1}\exp(-Q/RT) \quad (\alpha\sigma \le 0.8) \\ \dot{\varepsilon} &= A_2\exp(\beta\sigma)\exp(-Q/RT) \quad (\alpha\sigma \ge 1.2) \\ \dot{\varepsilon} &= A[\sinh(\alpha\sigma)]^n\exp(-Q/RT) \ (\alpha\sigma \text{ is arbitrary value}) \end{aligned} \tag{3.14}$$

where $A_1 = A\alpha^{n_1}$, $A_2 = \frac{A}{2^n}$, $\beta = \alpha n_1$.

In 1944, Zener and Hollomon proposed a temperature-compensated strain rate parameter [10], that is, the Z parameter to characterize the combined effect of deformation temperature and strain rate on flow stress. The expression is

$$Z = \dot{\varepsilon}\exp(Q/RT) \tag{3.15}$$

Many researchers have established the constitutive model of a variety of materials with the Z parameter, and the results show that the model can describe the flow behavior of the material very accurately [11].

Take the logarithm of both sides of Eq. (3.14) to get the following three expressions,

$$\ln\dot{\varepsilon} = \ln A_1 + n_1\ln\sigma - Q/RT \tag{3.16}$$

$$\ln\dot{\varepsilon} = \ln A_2 + \beta\sigma - Q/RT \tag{3.17}$$

$$\ln\dot{\varepsilon} = \ln A_2 + n\ln(\sinh(\alpha\sigma)) - Q/RT \tag{3.18}$$

It can be seen that when the temperature is constant, there is a linear relationship between $\ln\dot{\varepsilon}$ and $\ln\sigma$, $\ln\dot{\varepsilon}$ and σ, $\ln\dot{\varepsilon}$ and $\ln(\sinh(\alpha\sigma))$. The peak stress σ_P under various deformation conditions listed in Table 3.1 was taken as σ.

The relationship between each parameter σ-$\ln\dot{\varepsilon}$, $\ln\sigma$-$\ln\dot{\varepsilon}$, $\ln(\sinh(\alpha\sigma))$-$\ln\dot{\varepsilon}$ was numerically fitted, as shown in Fig. 3.3. The following material parameters can be obtained, $n_1 = 8.142825$, $\beta = 0.14561$, $\alpha = 0.017882$, $n = 5.913223$, $Q =$ 262.9543 kJ/mol. Then the relationship between lnZ and $\ln[\sinh(\ln(\sinh(\alpha\sigma)))]$ was obtained by linear fitting, as shown in Fig. 3.4, so $n = 5.65048$, $\ln A = 42.7767$, $A =$ 3.78168E+18. Substituting the parameters into the model, the constitutive equation of 2024 aluminum alloy can be obtained:

Table 3.1 Peak stress under different deformation conditions

T/°C	$\dot{\varepsilon}$/s^{-1}	σ_P/MPa
300	0.001	102.95
	0.01	119.46
	0.1	134.57
	1	146.3
350	0.001	69.01
	0.01	80.32
	0.1	93.86
	1	112.26
400	0.001	25.41
	0.01	44.28
	0.1	62.39
	1	75.6
450	0.001	15.4
	0.01	32.24
	0.1	53.65
	1	66.73

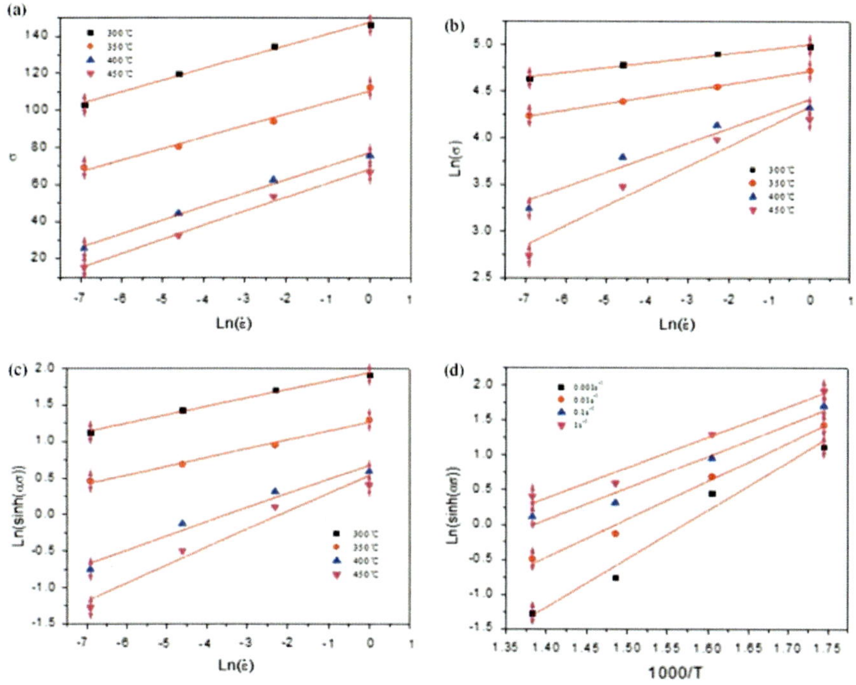

Fig. 3.3 Relationship between the parameters: **a** σ-ln $\dot{\varepsilon}$, **b** ln σ-ln $\dot{\varepsilon}$, **c** ln(sinh($\alpha\sigma$))-ln $\dot{\varepsilon}$, **d** ln(sinh($\alpha\sigma$))-1/T

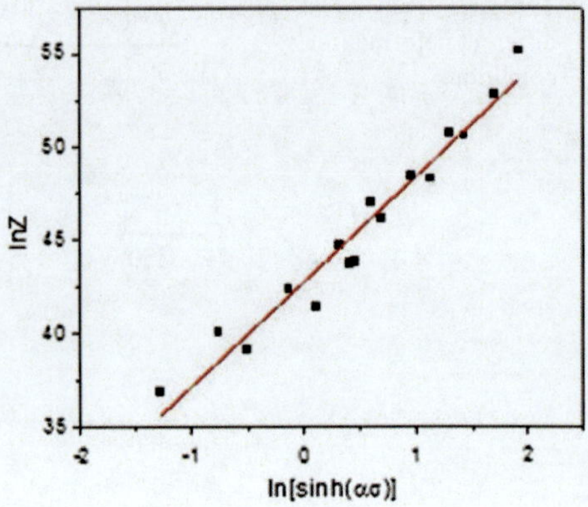

Fig. 3.4 Relationship between lnZ and ln[sinh($\alpha\sigma$)]

$$\dot{\varepsilon} = 3.78168 \times 10^{18}[\sinh(0.017882\sigma)]^{5.913223} \times \exp\left(-2.629543 \times \frac{10^5}{RT}\right) \quad (3.19)$$

3.4.2 Dynamic Recrystallization Modeling of 2024 Aluminum Alloy

(1) Deformation experiment and grain measurement

During the hot deformation, the evolution of the microstructure is affected by temperature, strain rate and strain. In order to establish the relationship between the grain size of the 2024 aluminum alloy and the temperature, strain rate and strain, hot compression experiments are also required. The experimental process was the same as that in above section. The compressed samples were cut into microstructure observation specimens along the direction parallel to the compression axis. The preparation steps were: cutting → mounting → grinding → mechanical polishing → electrolytic polishing. Because the aluminum alloy is soft, the specimens were polished on 800#, 1000#, 2000# sandpaper in sequence until there were no obvious scratches. The diamond polishing paste with grain size of 5, 2.5 and 1 μm were used successively during the mechanical polishing. The electrolyte mixed with 10% perchloric acid and 90% absolute ethanol was employed for electrolytic polishing. The electrolyte temperature was −20 °C. During polishing, the voltage was 20 V, and the electrolysis time was about 20 s. After electrolytic polishing, the specimens were observed immediately.

Because the microscope cannot directly measure the grain size, the electron backscattered diffraction (EBSD) technique was used to measure the grain size. The principle is that the electron beam bombards the observation surface in a certain direction, and diffraction occurs between the grains or on the crystal lattice surface inside the grains, resulting in a Kikuchi pattern. Thus, the misorientation between different grains in the observed specimen can be obtained. Then the morphology, size and distribution of high-angle grain boundaries or low-angle grain boundaries can be reconstructed by imaging technology. Compared with the scanning electron microscope analysis technology, EBSD can obtain the distribution of different grain orientations, and it is easier to obtain the orientation of the recrystallized grains, size, and proportions for the study of the microstructure evolution. The equipment used in the EBSD tests was the Oxford Channel 5 EBSD System.

The EBSD results of raw materials and compressed specimens are shown in Fig. 3.5. The thick lines in the figure represent high-angle grain boundaries with misorientation greater than 15°, and the thin lines represent low-angle grain boundaries with misorientation between 2° and 15°. It can be seen that the original microstructure of 2024 aluminum alloy is long-axis grains, with relatively straight grain boundaries, and there are almost no low-angle grain boundaries in the grains. The microstructure of the compressed specimen shows that many grain boundaries appear to be serrated, and some fine recrystallized grains are distributed along them. The serrated characteristics of deformed grain boundaries indicate that the dynamic recrystallization mechanism is mainly grain boundary migration. At the initial stage of deformation, the migration of grain boundaries causes dislocations to accumulate at the grain boundaries. Meanwhile, the unbalanced dislocation density near the original grain boundaries causes stress concentration and the grain boundaries to bow. The dislocation density in the migration region of the grain boundaries is

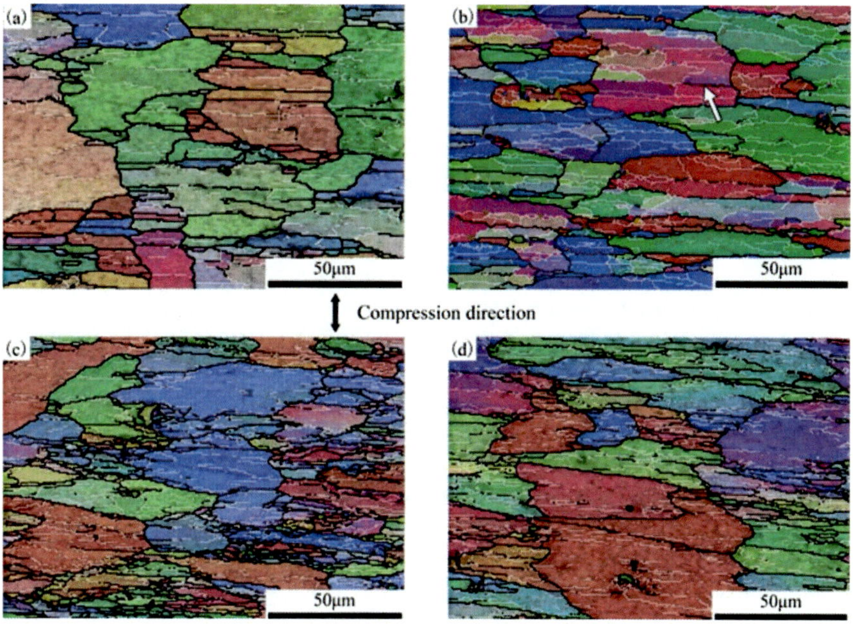

Fig. 3.5 EBSD diagram of compressed specimens: **a** 450 °C, 0.001 s^{-1}, strain 0.3, **b** 450 °C, 0.001 s^{-1}, strain 0.7, **c** 350 °C, 0.001 s^{-1}, strain 0.7, **d** 450 °C, 0.1 s^{-1}, strain 0.7 [12]

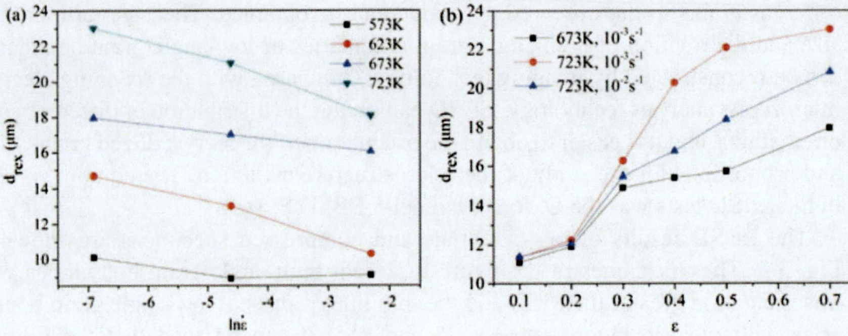

Fig. 3.6 The average size of recrystallized grains of the compressed specimens: **a** the influence of temperature and strain rate, **b** the influence of strain [12]

redistributed to form low-angle grain boundaries. As the deformation continues, some low-angle grain boundaries absorb dislocations and gradually transform into high-angle grain boundaries, such as the grain boundaries pointed by the arrow in Fig. 3.5b, thereby new grains in the original grains are generated. The low-angle grain boundary extended to the inside and connects with other low-angle grain boundaries, which will also divide the original grains into fine grains [12].

The comparison of microstructure in Fig. 3.5 shows that when the temperature is increased from 350 to 450 °C, the proportion of recrystallized grains and the size of the recrystallized grains both increase. When the strain rate is 0.1 s^{-1}, only very few recrystallized grains are generated. As the strain increases, the low-angle grain boundaries and fine recrystallized grains increase significantly. The recrystallized grain size measured by the EBSD system is shown in Fig. 3.6. The size of recrystallized grains d_{rex} increases with the decrease of strain rate and temperature, and the increase of strain. As the temperature increases, the diffusion activation energy of atoms increases, resulting in a faster diffusion rate of atoms and easier movement of dislocations at high temperatures, which improves the ability of grain boundary migration. Under the condition of low strain rate, the stress concentration can be fully released within a certain time, so that the dislocations can be fully diffused, and the growth of recrystallized grains can be promoted.

(2) Critical strain of dynamic recrystallization

The theory of dynamic recrystallization believes that dynamic recrystallization occurs before the stress reaches its peak. According to the theory of work hardening, the change of the work hardening rate θ with the stress σ is shown in Fig. 3.7, which can generally be divided into 5 stages [13]. The first stage is called the easy-slip stage. Strong dislocation slip and multiplication ability results in a very low work hardening rate. The second stage is called the linear hardening stage. The density of dislocations increases rapidly, and the accumulation and entanglement of dislocations occur. As the degree of deformation increases, the flow stress increases significantly. The third stage is the dynamic recovery hardening stage. The total dislocation density is still

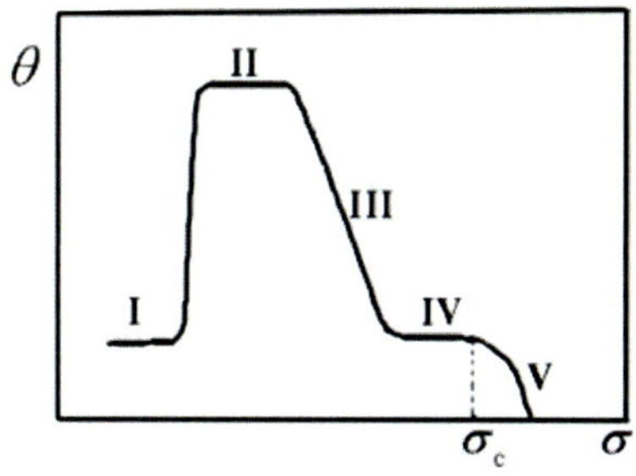

Fig. 3.7 Schematic of the relationship between work hardening rate θ and stress σ

increasing due to the simultaneous multiplication and annihilation of dislocations, but there is also a softening effect caused by dynamic recovery. The work hardening rate decreases proportionally with the increase of stress. The fourth stage is the large strain hardening stage. The cellular structure formed by the movement of dislocations continuously absorbs dislocations and forms sub-grains, resulting in the work hardening rate reaching a stable value. At fifth stage, dynamic recrystallization occurs, and the work hardening rate begins to decrease again. The stress corresponding to the inflection point between stage IV and stage V is the dynamic recrystallization critical stress σ_c.

To obtain the work hardening rate curve of 2024 aluminum alloy, it is necessary to calculate the slope of each point on the stress–strain curve. However, the data points collected by the testing machine were fluctuating. A ninth degree polynomial was used to fit a smooth and continuous stress–strain curve. Differential of the polynomial equation obtained by fitting was performed to obtain the θ value, and then the θ-σ curve was drawn, as shown in Fig. 3.8.

The θ-σ curve was differentiated to obtain the $-\partial\theta/\partial\sigma$-$\sigma$ curve, as shown in Fig. 3.9. The stress at the minimum inflection point of the curve is the critical stress for dynamic recrystallization, and then the corresponding strain is the critical strain. The critical strain under each deformation condition is listed in Table 3.2. It can be seen that as the temperature increases, the critical strain decreases, indicating that the increase in temperature is conducive to the occurrence of dynamic recrystallization of 2024 aluminum alloy. As the strain rate increases, the critical strain increases, and the beginning of dynamic recrystallization is later.

There is a linear relationship between the critical strain and the peak strain, which can be expressed as $\varepsilon_c = \alpha\varepsilon_p$. Therefore, $\alpha = 0.38593$ can be obtained. According to the stress–strain curve, the peak strain is related to temperature, strain rate, initial grains, etc. The peak strain can be expressed as [14]

$$\varepsilon_p = a_1 d_0^{n_1} \dot{\varepsilon}^{m_1} \exp\left(\frac{Q_1}{RT}\right) \tag{3.20}$$

where d_0 is the initial grain size; Q_1 is the activation energy for recrystallization; $\dot{\varepsilon}$ is the strain rate; R is the gas parameter; a_1, n_1, and m_1 are the material constants.

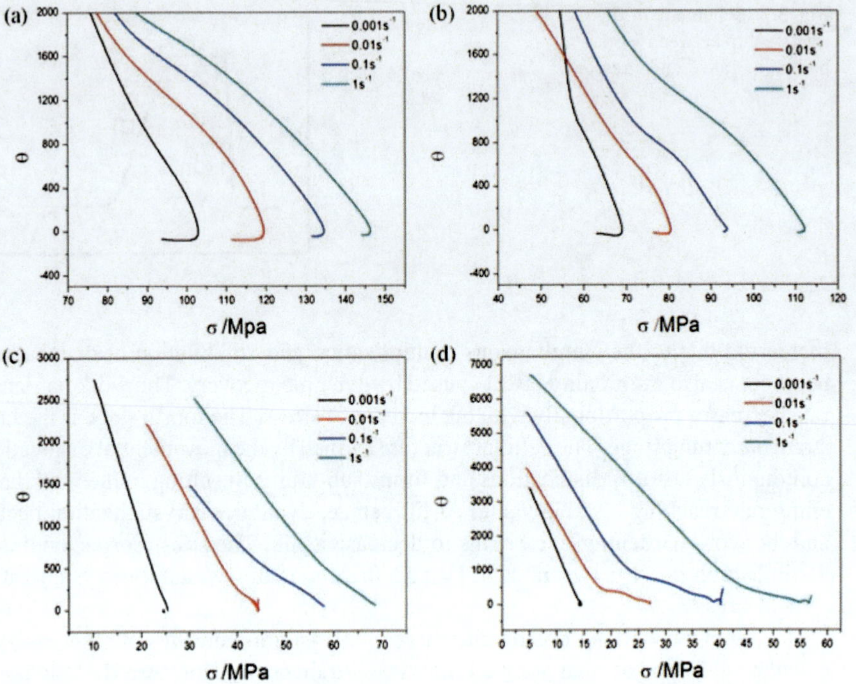

Fig. 3.8 The relationship between work hardening rate θ and stress σ: **a** $T = 300$ °C; **b** $T = 350$ °C; **c** $T = 400$ °C; **d** $T = 450$ °C

The initial grain size of the 2024 aluminum alloy used in this study was 25 μm, so $a_1 d_0^{n_1}$ can be replaced by the constant A_1. Take the logarithm of both sides of the Eq. (3.20):

$$\ln \varepsilon_p = \ln A_1 + m_1 \ln \dot{\varepsilon} + \frac{Q_1}{RT} \tag{3.21}$$

Under constant temperature conditions, a numerical regression analysis was performed on $\ln \varepsilon_p$ and $\ln \dot{\varepsilon}$. The relationship between $\ln\varepsilon_p$ and $\ln\dot{\varepsilon}$ can be obtained, as shown in Fig. 3.10a. Take the average of the slope of each fitting line, $m_1 = 0.1371225$. Under the condition of constant strain rate, linear regression was performed on $\ln\varepsilon_p$ and $1/T$. The relationship between $\ln \varepsilon_p$ and $1000/T$ can be obtained, as shown in Fig. 3.10b. The average value of the slope of each line is the value of $Q_1/1000$, so $Q_1 = 16{,}198.5981$. The intercept of the fitting line is $\ln A_1 + m_1 \ln \dot{\varepsilon}$, so that $A_1 = 0.008535$. Finally, the obtained critical strain of dynamic recrystallization is:

$$\varepsilon_c = 0.38593\varepsilon_p \tag{3.22}$$

$$\varepsilon_p = 0.008535 \times \dot{\varepsilon}^{0.1371225} \times \exp(16198.5981/RT) \tag{3.23}$$

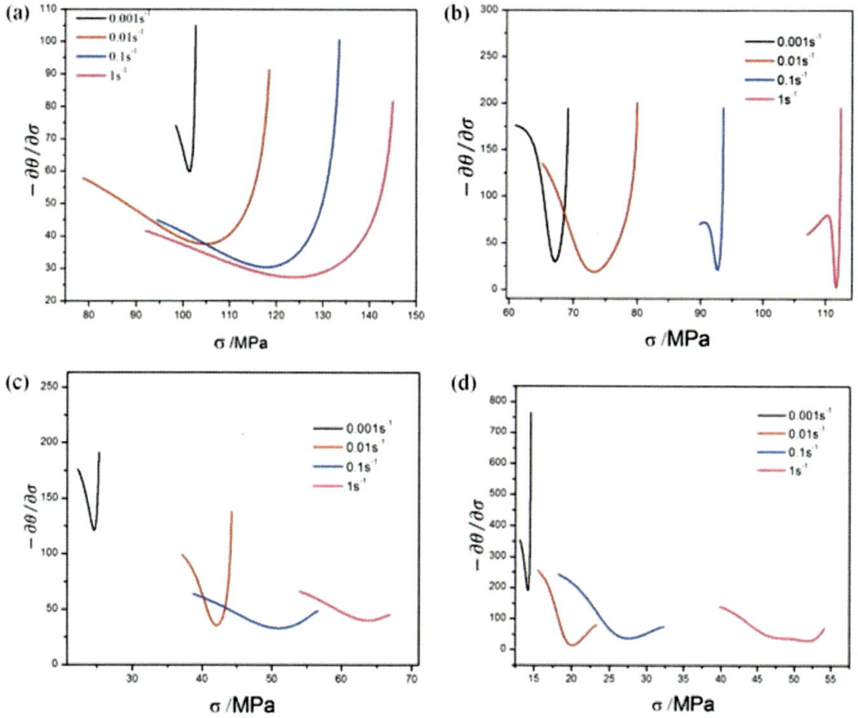

Fig. 3.9 The relationship between $-\partial\theta/\partial\sigma$ and stress σ **a** T = 300 °C; **b** T = 350 °C; **c** T = 400 °C; **d** T = 450 °C

Table 3.2 Critical strain of dynamic recrystallization under different deformation conditions

T /°C	$\dot{\varepsilon}$/s^{-1}			
	0.001	0.01	0.1	1
300	0.03402	0.03801	0.04634	0.0465
350	0.02197	0.02985	0.03384	0.03569
400	0.01962	0.02007	0.02671	0.03646
450	0.01164	0.01587	0.01441	0.01705

(3) Dynamic recrystallization volume fraction

The dynamic recrystallization volume fraction characterizes the degree of dynamic recrystallization. Assuming that the grains are spherical, according to the Avrami equation, the volume fraction of dynamic recrystallization X_{drex} can be expressed as

$$X_{\text{drex}} = 1 - \exp\left(-\text{B}t^n\right) \tag{3.24}$$

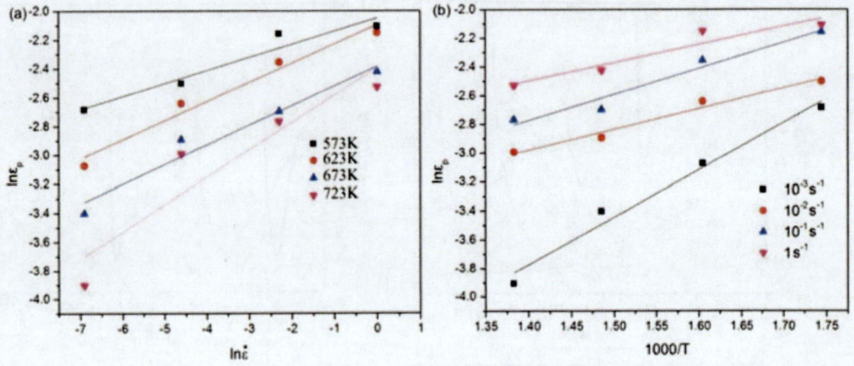

Fig. 3.10 The relationship between the parameters; **a** $\ln\varepsilon_p$-$\ln\dot{\varepsilon}$; **b** $\ln\varepsilon_p$-1000/*T*

where *t* is the deformation time; *B* is a constant. The strain can be regarded as a function of time, then Eq. (3.24) can be transformed into

$$X_{\text{drex}} = 1 - \exp\left\{-\beta_{\text{d}}[(\varepsilon - \varepsilon_{\text{c}})/\varepsilon_{0.5}]^{k_{\text{d}}}\right\} \tag{3.25}$$

where $\varepsilon_{0.5}$ is the strain when the volume fraction of dynamic recrystallized grains is 50%; β_{d} and k_{d} are the material constant.

There are energy method, quantitative metallographic method and stress–strain curve method to determine the recrystallization volume fraction. Because it is difficult to measure the energy stored inside the material, the energy method is generally not used. The quantitative metallographic method requires more work than stress–strain curve method, so this book introduces the stress–strain curve method.

For the stress–strain curve, if there is no dynamic recrystallization, the change of the stress–strain curve can be simply divided into two stages: one is the work hardening stage, where the stress increases rapidly; the other is the steady-state stage. If dynamic recrystallization occurs, the stress will decrease to a steady state after reaching a peak due to softening. The recrystallization degree of different strains can be expressed by the degree of softening. Therefore, according to the stress–strain curve, the dynamic recrystallization volume fraction can be expressed as

$$X_{\text{drex}} = \frac{\sigma_{\text{REC}} - \sigma_{\text{DRX}}}{\sigma_{\text{sat}} - \sigma_{\text{ss}}} \tag{3.26}$$

where σ_{REC} is the instantaneous flow stress of dynamic recovery; σ_{sat} is the steady state flow stress of dynamic recovery; σ_{DRX} is the instantaneous flow stress of dynamic recrystallization; σ_{ss} is the steady state flow stress of dynamic recrystallization.

The mathematical model of dynamic recovery flow stress can be expressed as [15]

$$\sigma = \left(\sigma_{\mathrm{sat}}^2 - \left(\sigma_{\mathrm{sat}}^2 - \sigma_0^2\right)\exp(-r\varepsilon)\right)^{0.5} \tag{3.27}$$

where r is the dynamic recovery rate, σ_0 is the initial stress. By derivation of flow stress to strain, the following relationship can be obtained.

$$\sigma\frac{\mathrm{d}\sigma}{\mathrm{d}\varepsilon} = 0.5r\sigma_{\mathrm{sat}}^2 - 0.5r\sigma^2 = \theta\cdot\sigma \tag{3.28}$$

where $-0.5r$ is the slope of the straight line segment of the $\theta\cdot\sigma$-σ^2 curve. The σ_{sat} is the stress value corresponding to the intersects point where the tangent line made based on the θ-σ curve passing through the critical point of dynamic recrystallization and $\theta = 0$ line.

Using the above method, the dynamic recrystallization volume fraction under different deformation conditions can be obtained, and the strain when the recrystallization volume fraction is 50% under certain conditions can also be obtained. The $\varepsilon_{0.5}$ is related to deformation temperature, strain rate and initial grain size, and can be expressed as

$$\varepsilon_{0.5} = a_2 d_0^{h_2}\dot{\varepsilon}^{m_2}\exp\left(\frac{Q_2}{RT}\right) \tag{3.29}$$

where a_2, h_2 and m_2 are the material constants; Q_2 is the activation energy.

By taking the logarithm of both sides, Eq. (3.29) can be transformed into

$$\ln\varepsilon_{0.5} = \ln\left(a_2 d_0^{h_2}\right) + m_2\ln\dot{\varepsilon} + \frac{Q_2}{RT} \tag{3.30}$$

By establishing the relationship of $\ln\varepsilon_{0.5}$-$\ln\dot{\varepsilon}$ and $\ln\varepsilon_{0.5}$-$1/T$, as shown in Fig. 3.11, the following parameters can be obtained: $m_2 = 0.07191$, $Q_2 = 28{,}802.178$; $\ln\varepsilon_{0.5} = 0.0023516$.

By transforming Eq. (3.30), the following equation can be obtained

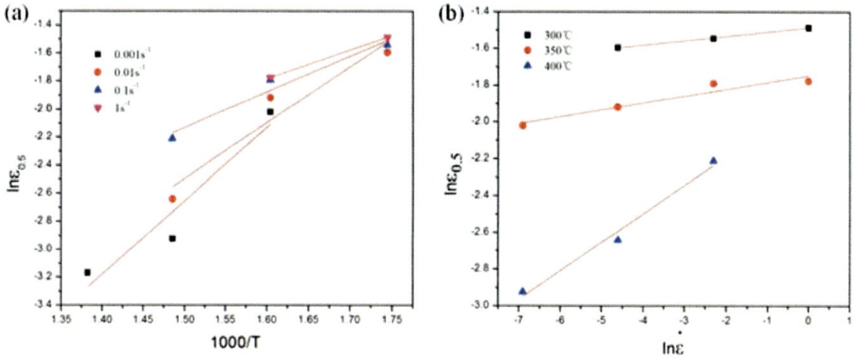

Fig. 3.11 The relationship between the parameters: **a** $\ln\dot{\varepsilon}$-$1/T$; **b** $\ln\varepsilon_{0.5}$-$\ln\dot{\varepsilon}$

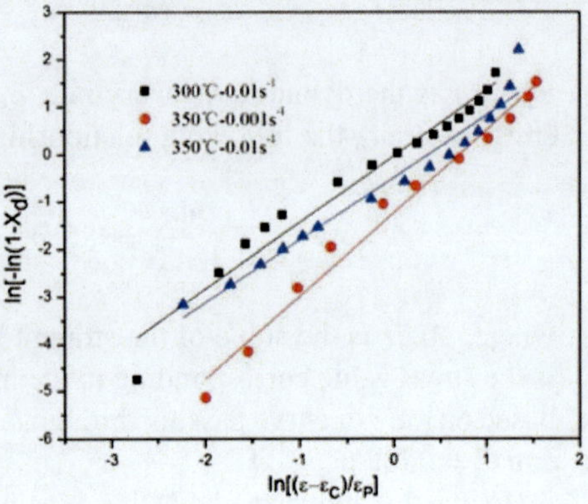

Fig. 3.12 The relationship between $\ln[-\ln(1-X_d)]$ and $\ln[(\varepsilon-\varepsilon_c)/\varepsilon_{0.5}]$

$$\ln[-\ln(1-X_d)] = \ln\beta_d + k_d\ln[(\varepsilon-\varepsilon_c)/\varepsilon_{0.5}] \tag{3.31}$$

By establishing the relationship of $\ln[-\ln(1-X_d)]$ and $\ln[(\varepsilon-\varepsilon_c)/\varepsilon_{0.5}]$, as shown in Fig. 3.12, the slope and intercept of each line were obtained according to the fitting parameters, thereby $\beta_d = 0.55265$, $k_d = 1.48793$. Finally, the dynamic recrystallization volume fraction is expressed as

$$X_{drex} = 1 - \exp\{-0.55265[(\varepsilon-\varepsilon_c)/\varepsilon_{0.5}]^{1.48793}\} \tag{3.32}$$

$$\varepsilon_{0.5} = 0.00023516 \times \dot{\varepsilon}^{0.07191}\exp\left(\frac{28802.178}{RT}\right) \tag{3.33}$$

(4) Average grain size during dynamic recrystallization

The average grain size during dynamic recrystallization is related to the deformation conditions and the internal factors of the material. It can be expressed as an equation about the initial average grain size, recrystallized grain size and recrystallization volume fraction, namely

$$d_{avg} = d_{rex}X_{drex} + d_0(1-X_{drex}) \tag{3.34}$$

In this study, the d_0 was 25 μm. The recrystallization volume fraction X_{drex} has been expressed by Eq. (3.32). Therefore, only the mathematical relationship between the recrystallized grain size and temperature, strain rate and strain needed to be established. The equation can be expressed as

$$d_{rex} = a_3 d_0^{h_3}\varepsilon^{n_3}\dot{\varepsilon}^{m_3}\exp(Q_3/RT) \tag{3.35}$$

where a_3, h_3 and m_3 are regression coefficients; Q_3 is activation energy.

Take the logarithm of both sides of the equation,

$$\ln d_{rex} = \ln a_3 d_0^{h_3} + n_3 \ln \varepsilon + m_3 \ln \dot{\varepsilon} + Q_3/RT \tag{3.36}$$

By establishing the relationship of $\ln d_{drex}$ with $\ln \varepsilon$, $\ln \dot{\varepsilon}$ and $1/T$, the values of parameters were obtaind: $Q_3 = -18{,}630.33165$, $m_3 = -0.05805$, $n_3 = 0.24115$, $a_3 d_0^{h_3} = 379.2568$. Therefore, the average grain size during dynamic recrystallization is as follows.

$$d_{\text{rex}} = 379.2568 \times \varepsilon^{0.24115} \dot{\varepsilon}^{-0.05805} \exp\left(\frac{-18630.33165}{RT}\right) \tag{3.37}$$

References

1. Ji DS (2011) Study on the forming mechanism and application of hollow divided flow forging technology. Huazhong Unviersity of Science and Technology, Wuhan
2. Chenot JL (1992) Recent contributions to the finite element modeling of metal forming processes. J Mater Process Technol 34(1–4):9–18
3. Li J, You LH (1999) Coupling analysis of deformation and heat conduction during hot forging. Mech Res Appl 12(2):19–21
4. Jin JH (2009) Research on the key technology of near-net shaped forming for gears used in car. Huazhong Unviersity of Science and Technology, Wuhan
5. Zhao T (2015) Study on flow behavior and microstructure evolution behavior of 2024 aluminum alloy sheet under different stress states. Huazhong Unviersity of Science and Technology, Wuhan
6. Chen YL, Chen WZ, Hong LH, Fu GS (2008) Research status of flow stress model of aluminum alloy at elevated temperatures. Foundry Technol 29(9):1223–1226
7. Sellars CM, Tegart WJ (1966) On the mechanism of hot deformation. Acta Metall 14(9):1136–1138
8. Dai J, Li X, Li SQ, Wang KL (2014) Constitutive equation of titanium alloy TC21 deformation at high temperature. J Netshape Forming Eng 6:116–121
9. Jia YJ (2013) Experimental research and numerical simulation of hot deformation and dynamic recrystallization behavior of 7050. Chongqing University, Chongqing
10. Zener C, Hollomon JH (1944) Effect of strain-rate upon the plastic flow of steel. J Appl Phys 15(1):22–27
11. Li XW, Xiong BQ, Zhang YA, Hua C, Wang F, Zhu BH, Xiong YM (2008) Flow stress feature of novel Al-Zn-Mg-Cu alloy during hot compression deformation. Chinese J Rare Metals 32(5):552–557
12. Deng L, Zhou P, Wang XY, Jin JS, Zhao T (2018) Microstructure evolution and modeling of 2024 aluminum alloy sheets during hot deformation under different stress states. Met Mater Int 24(1):112–120
13. Estrin Y, Tóth LS, Molinari A (1998) A dislocation-based model for all hardening stages in large strain deformation. Acta Materialia 46(15):5509–5522
14. Deng L (2011) Applid fundamental study on precision forging of aluminum alloy. Huazhong University of Science and Technology, Wuhan
15. Jonas JJ, Quelennec X, Jiang L (2009) The Avrami kinetics of dynamic recrystallization. Acta Mater 57(9):2748–2756

Chapter 4
Precision Forging Technology for Long Shaft Parts

4.1 Overview

Long shaft forgings account for a large proportion of aluminum alloy forgings, among which aluminum alloy control arms for automobiles are the most typical [1]. It can be further divided into straight or curved forgings, such as the planar control arm, multi-directional control arm, branch-shaped control arm, multi-directional curved control arm, etc., and complex forgings, such as the wingspan control arm and herringbone control arm, as shown in Fig. 4.1.

During die forging of this kind of forgings, the axis of the billet is perpendicular to the direction of the forging force. Therefore, it can be approximately considered that the metal basically flows along the height and width in the plane perpendicular to the axis. The metal flow in the semicircular part at both ends of the forging is the same as that of the short-shaft axisymmetric parts.

The general die forging process of long shaft aluminum alloy forgings is: cutting → heating → stretching → bending (for curved shaft forgings) → pre-forging (for complex long shaft forgings) → final forging → trimming, punching → sizing. The following will focuses on the final forging, pre-forging and billet-making and die design, and a brief introduction to the trimming, punching and sizing process and die design.

4.2 Design of Final Forgings and Die Cavity

The long shaft aluminum alloy forging is difficult to form by closed precision forging without flash, so the flash must be considered in the design of final forgings. The final forging is completed by the final forging die cavity which is composed of a die cavity designed according to the hot forging drawing and a flash groove that is set around the parting surface.

L. Deng et al., *Precision Forging Technology and Equipment for Aluminum Alloy*, Springer Series in Advanced Manufacturing,
https://doi.org/10.1007/978-981-19-1828-5_4

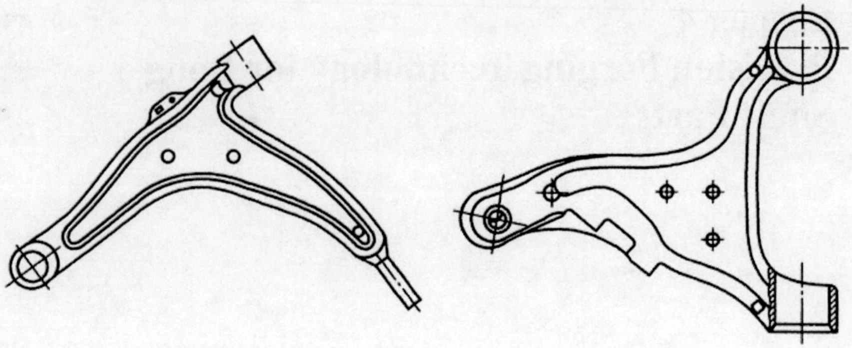

Fig. 4.1 Long shaft aluminum alloy forgings for automobiles

4.2.1 Design of Hot Forgings

The following points should be considered when designing.

(1) The hot forging drawing is designed according to the cold forging drawing, but the size on the hot forging drawing is larger than the corresponding size on the cold forging drawing. Theoretically, the size L considered the shrinkage rate is calculated as follows:

$$L = l(1 + \delta)$$

where l is the size of cold forging; δ is the shrinkage rate of the materials at the final forging temperature, 1% for aluminum alloy, and 0.8–0.9% for slender parts.

(2) Since the die cavity is easy to wear, the amount of wear should be added within the negative tolerance of the forging to increase the life of forging die.

(3) Since the shape of the forging cannot ensure the accurate positioning of the billet in the lower die or in the trimming die, the positioning block should be designed to the hot forging drawing. It can be removed during trimming or cutting.

(4) When the forging is easily deformed during trimming and punching, a certain amount of compensation should be properly considered in the deformed region of the hot forging to improve the qualification rate of the forging.

(5) The complex parts and high parts of the forging should be placed on the upper die. This is because lubricant residues is easy to accumulate in the deep of the lower die, resulting in underfilling of the forging. If these parts of the forging must be placed in the lower die, the size of these parts on the hot forging drawing should be increased.

(6) The positions and dimensions of the parting surface and the wad should be indicated on the hot forgings.

(7) The unmarked draft angle, fillet radius and shrinkage rate should be written on the hot forging drawing.
(8) In order to facilitate the processing and inspection of the die cavity, the height should be marked on the basis of the parting surface.

4.2.2 Design of Flash Groove

According to the deformation mechanism and flow characteristics of aluminum alloy forgings, the functions of flash groove are as follows:

(1) Creating a large enough horizontal metal flow resistance, and promoting the metal to flow in the longitudinal direction to fill the die cavity.
(2) To accommodate the excess metal on the billet, which can compensate and adjust the volume change caused by the fluctuation of the billet volume and the wear of the die cavity.
(3) Cushioning effect for impact die forging equipment, which can avoid the upper die and lower die from hitting, thereby preventing collapse or cracking of the parting surface.

The flash groove is composed of a bridge part and a bin part, as shown in Fig. 4.2. There are three basic structural forms [2]:

(1) Form I is the most widely used flash groove. Its advantage is that the bridge is set on the upper die, and not easy to wear and collapse due to less heating by the billet.
(2) Form II is used for asymmetric forgings in height direction. When the forging process is carried on hammers and screw presses, the complex shape of the forging is often placed on the upper die to facilitate the filling and forming.

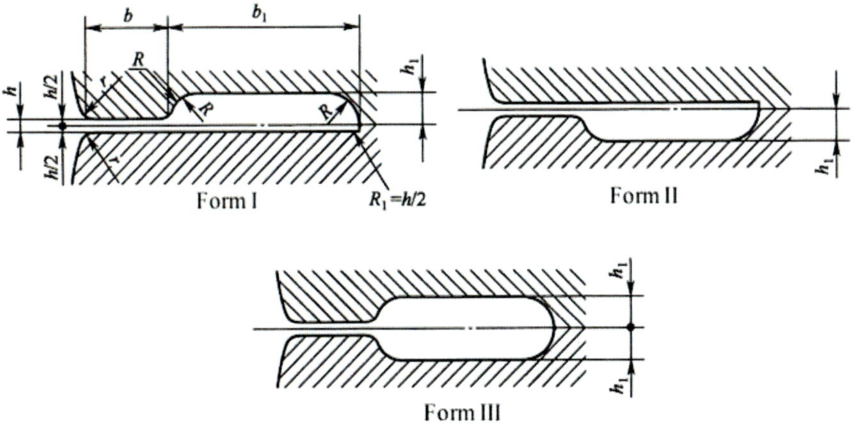

Fig. 4.2 Three structural forms of flash groove

But when trimming, the forging should be turned over 180° to simplify the punch of the trimming die. For this reason, the bridge is set on the lower die. In addition, when the entire forging is formed in the lower die, in order to simplify the shape of the upper die, this form of flash groove should also be used.

(3) Form III is suitable for aluminum alloy forgings with complex shapes to accommodate more flash metal.

The main dimensions of the flash groove are the height h_f and width b. When b does not change, the resistance decreases with the increase of h_f. When h_f does not change, the resistance increases with the increase of b.

The projected area of the forging on the parting surface is not only the main basis for selecting the flash groove size, but also for selecting the tonnage of the equipment. Therefore, the flash groove size is usually selected according to the tonnage of the forging equipment. The dimensions of the flash grooves of the die used on the common die forging hammer, screw press and hot die forging press are listed in Tables 4.1, 4.2 and 4.3, respectively.

In addition to using the tonnage method to determine the size of the flash groove, the following equation can also be used to determine the height of the bridge.

$$h_f = 0.015\sqrt{A_P} \quad (4.1)$$

where A_p is the projected area of the forging on the parting surface.

Table 4.1 Flash groove size on die forging hammer

Hammer tonnage/t	h_f	h_1	b	b_1	Remarks
1	1–1.6	4	8	25	Gear lock $b_1 = 30$
1.5	1.6–2	4	8	25–30	
2	2	4	10	30–35	Gear lock $b_1 = 40$
3	3	5	12	30–40	Gear lock $b_1 = 45$
5	3	6 × 2	12	50	Gear lock $b_1 = 55$
10	5	6 × 2	16	50	
16	7	8 × 2	18	65	

Table 4.2 Flash groove size on screw press

Press tonnage/kN	h_f	h_1	b	b_1	r	R
≤1600	1.2	4	6	25	1.5	4
1600–4000	1.5	4	8	30	2.0	4
4000–6300	2.0	5	8	35	2.0	5
6300–10,000	2.5	6	10	35	2.5	6
10,000–25,000	3.0	7	12	40	3.5	7

Table 4.3 Flash groove size on hot die forging press

Size/mm	Press tonnage/kN								
	10,000	16,000	20,000	25,000	31,500	40,000	63,000	80,000	120,000
h_f	2	2	3	4	5	5	6	6	8
b	10	10	10	12	15	15	20	20	24
B	10	10	10	10	10	10	10	12	18
L	40	40	40	50	50	50	60	60	60
r_1	1	1	1.5	1.5	2	2	2.5	2.5	3
r_2	2	2	2	2	3	3	4	4	4

The values in Tables 4.1, 4.2 and 4.3 is the projected area of the forging on the parting surface. When the shape of the forging is complex, in order to increase the flash resistance and ensure the forming of the forging, the value of b should be appropriately increased. When the shape of the forging is relatively simple, in order to reduce the wear of the die while ensuring the shape of the forging, the value of h_f can be appropriately increased or the value b decreased.

When die forging process is carried out on a hot die forging press, the height of the forging is guaranteed by the stroke of the press. Therefore, when the slider is at the bottom, there must be a certain gap between the upper die and lower die to adjust the closing height of the die and offset part of the elastic deformation of the press, so as to ensure the dimensional accuracy of the forging in the height direction. The gap between the upper die and lower die can also prevent the press from jam. The size of the gap is determined by the height of the flash groove. When the distance between the bin part of flash groove and the edge of the die is less than 20–25 mm, the groove can be directly penetrated to the edge of the die.

4.3 Small Flash Precision Forging Process

4.3.1 *The Influence of Flash Bridge Size on the Stress*

The approximate theoretical analysis based on the change of stress σ_z in the die cavity shows that when the width of the flash bridge b decreases, the maximum compressive stress at the center of the die cavity increases. The change of radial stress σ_r is similar to this. When the height of the flash bridge h_f is reduced, the change of the compressive stress in the die cavity is the same as the change of the width of the flash bridge. As explained above, with the decrease of the height of the flash bridge or the increase of the width, the three-directional compressive stress state in the die cavity during precision forging is more servere, which is beneficial to the longitudinal flow of metal to make the forging completely formed.

The shape and size of the flash groove is related to the shape and size of the forging, and even is related to the volume and shape of the billet before the final forging. The reasonable shape and size of the flash groove should not only ensure that the forging completely formed and can accommodate excess metal, but also should make the forging die has a long service life.

In order to generate enough radial resistance in the flash groove, accommodate the excess metal, and facilitate the removal of flash, the height of the flash bridge should be smaller and the width should be larger, and the height and width of the flash bin should be appropriate.

4.3.2 The Relationship Between Flash Bridge Size and Flash Volume

The relationship between the ratio of the width to the height of the flash bridge b/h_f and the volume of the flash V_f and the forging force P are shown in Fig. 4.3. It can be seen that with the increase of b/h_f the volume of the flash decreases, and the forging force increases. When $4 < b/h_f < 6$, the two curves intersect, and the value of b/h_f at the intersection point is the best value [3].

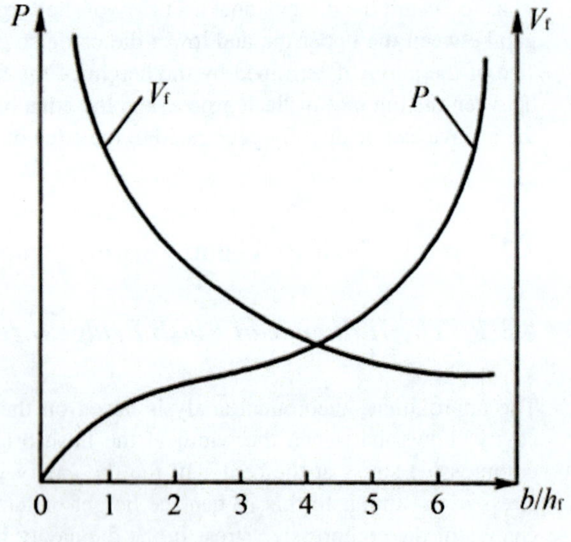

Fig. 4.3 The relationship between forging force, flash volume and flash bridge size

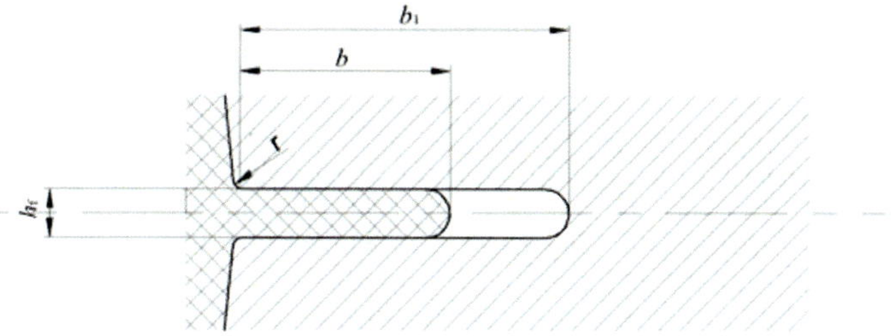

Fig. 4.4 Diagram of small flash groove and small flash

4.3.3 Optimized Design of Small Flash Groove

Based on the above analysis, it is possible to eliminate the bin part of the flash groove on the traditional final forging die cavity. A new flash groove structure with only bridge can be designed, as shown in Fig. 4.4. The process test and production practice show that when the flash width-to-height ratio on the forging is $b/h_{\text{f}} \geq 6$, the forging force P increases, but the flash volume V_{f} can be reduced by about 60%, which can effectively improve the material utilization. Because a circle of thin and flat flash is formed around the parting surface of the forging, it is also called flat thin flash or small flash.

The author has developed some hot precision forging processes of flat thin flash for complex long shaft forgings. Compared with the traditional die forging process, the material utilization has been increased from 73% to more than 90%, and the quality of forgings is better. Therefore, the development of the small flash forging process to realize the precision forging of long shaft parts has better economic benefits.

4.4 Design of Pre-forging Process and Die Cavity

4.4.1 The Role of Pre-forging Process

The role of pre-forging includes two aspects. One is to further distribute metal volume after billet-making to ensure that high-quality forgings without defects can be obtained during the final forging. The other is to reduce the wear of the final forging die and improve its service life [4].

When the forging structure is complex or has a rib-web section, such as connecting rods, herringbone steering control arms and wingspan-shaped components, the final forging can only be formed smoothly through pre-forging steps. Even if the shape of the high-strength aluminum alloy forging is not complex, but it has poor plasticity, so the pre-forging setup also should be used.

4.4.2 Design of Pre-forging Die Cavity

The pre-forging die cavity is designed based on the final forging die cavity, but there are differences between the two. The key points are as follows:

(1) During final forging process, the billet should be mainly deformed by upsetting as much as possible. Therefore, the height of the pre-forging die cavity should be 2–5 mm larger than that of the final forging die cavity, and the width should be 1–2 mm smaller. In addition, the pre-forging die cavity is not provided with flash groove, so the cross-sectional area of the pre-forging die cavity should be slightly larger than the corresponding cross-sectional area of the final forging die cavity. The difference between the two is shown in Fig. 4.5. The left is the pre-forging, and the right is the final forging.

(2) The draft angle of the pre-forging die cavity is generally to be the same as that of the final forging die cavity. A reasonable design scheme for extrusion deformation mode is shown in Fig. 4.6, that is, the height of the rib $h' = (0.8-0.9)h$, If the ratio of height-to-width of the rib h/a are relatively large,

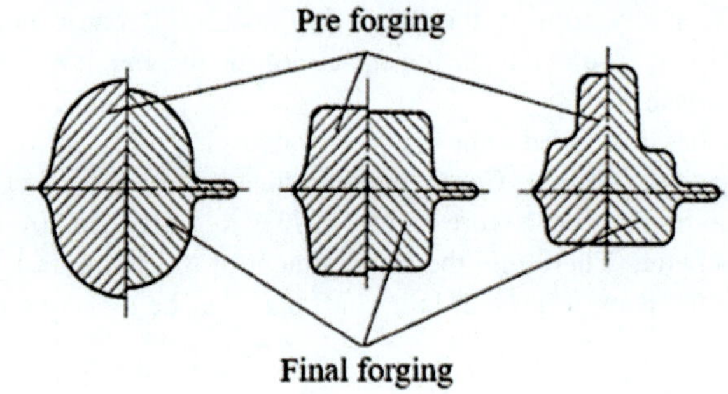

Fig. 4.5 The difference between the pre-forged shape and the final forged shape

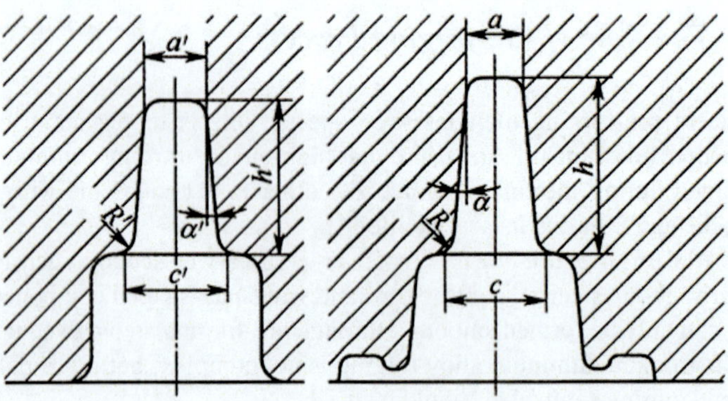

Fig. 4.6 Dimensional relationship between pre-forging die cavity and final forging die cavity

the coefficient takes the small value; otherwise, the coefficient takes the large value. Take the draft angle $a' = a$, there is $c' < c$. In order to make the forming of the ribes smoothly during the final forging, the cross-sectional area of the ribs of the pre-forging part should be not less than the corresponding area of the final forging, so the fillet radius of the bottom R' should be appropriately increased. In addition, the increase of R' is beneficial to the metal flow during the pre-forging.

(3) The fillet radius in the pre-forging die cavity should be larger than that of the final forging die cavity. Its purpose is to reduce metal flow resistance, promote the forming of the ribs. The fillet radius R' can be calculated as follows:

$$R' = R + c \tag{4.2}$$

where R is the fillet radius on the corresponding part of the final forging die cavity; c is the coefficient, if the depth of the final forging die $h < 10$, then $c = 2$, if $h = 10-25$, then $c = 3$, if $h = 25-50$, then $c = 4$, if $h > 50$, $c = 5$. The fillet radius of the pre-forging die cavity at the corners of the horizontal plane should be appropriately increased to make the billet deform smoothly to prevent folding during pre-forging and final forging.

(4) For the pre-forging die cavity with branches, in order to facilitate the flow of metal into the branch cavity, the shape of the branch of the pre-forging die cavity can be simplified, and the fillet radius at the junction with the branch should be appropriately increased. A resistance groove is set on the parting surface to increase the lateral resistance of the metal flowing to the flash groove during pre-forging, as shown in Fig. 4.7.

(5) For the pre-forging die cavity of the fork-shaped forging, when the distance between the branches is not large, the splitting boss must be used in the pre-forging. During pre-forging, the metal is split and squeezed to both sides by the splitting boss, and flows into the die cavity of the branch. In general, the

Fig. 4.7 Pre-forging die cavity with branches

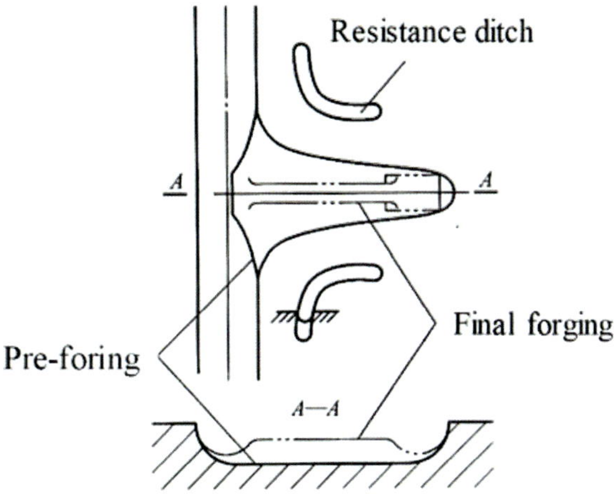

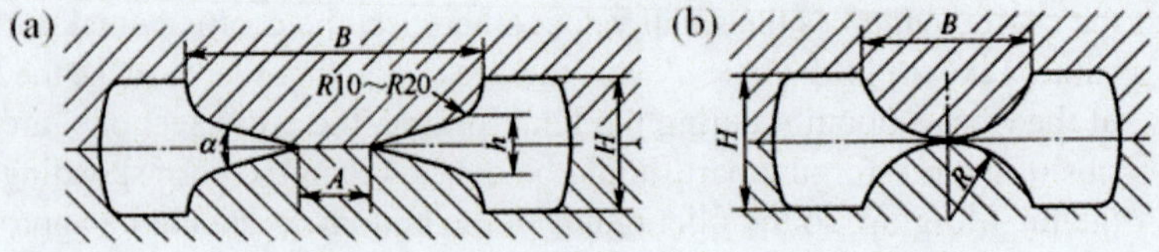

Fig. 4.8 The form of splitting boss: **a** with wide forks, **b** with narrow forks

form of Fig. 4.8a is used. When the fork-shaped part is narrow, the form of Fig. 4.8b can be used.

(6) For the pre-forging die cavity of rib-web section, such as various connecting rods, it is usually to design the pre-forging die cavity according to the relative height of the ribs and employ the different pre-forging methods shown in Fig. 4.9.

When $h \leq 2b$, the width of the pre-forging die cavity is $B' = B - (2\text{--}3)$, where B is the width of the final forging die. The height of the pre-forging die cavity h' is determined according to that the cross-sectional area of the pre-forging die cavity equal to the sum of the cross-sectional area of the final forging die cavity and the cross-sectional area of the flash, as shown in Fig. 4.10.

When $h > 2b$, $B' = B - (2\text{--}3)$mm. The determination method of h' is to assume that the pre-forging die cavity is a trapezoidal section and obtain H', then after the $x = (H - H')/4$ is calculated, the shape of the pre-forging die cavity is obtained by using a circular arc line, and $h' = H' + x$ can be obtained. When designing, the reduced area of the web part A_2 should be equal to the increased area of the ribs A_1, as shown in Fig. 4.10b.

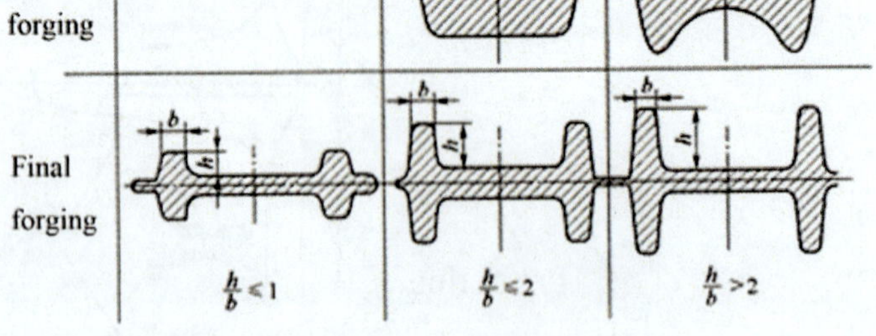

Fig. 4.9 Different pre-forging methods for rib-web section forgings

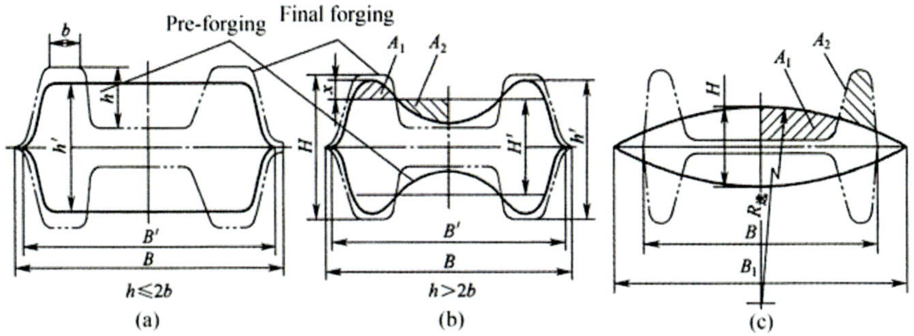

Fig. 4.10 Design of pre-forging die cavity for rib-web section forgings: **a** when $h \le 2b$, the pre-forged section is rectangular; **b** when $h > 2b$, the pre-forged section is saddle-shaped; **c** the pre-forged section of the high-rib-thin-web forging is elliptical

Figure 4.10c is a new design method. Firstly, the width of the pre-forging die cavity B_1 is determined according to the width of the final forging die cavity B, i.e. $B_1 = B + (10–20)$. Then an arc is drawn to make the area $A_1 = (1.0–1.1)A_2$. The purpose of $B_1 > B$ is to force the metal to form flash firstly, and then fill the cavity during the final forging. The pre-forging die cavity designed by the new method can avoid the flow line defects of vortex and penetrating ribs, and improve the quality of forgings.

4.5 Billet-Making Process for Long Shaft Forgings

4.5.1 Selection of Billet-Making Process

There are four types of long shaft forgings: straight long shaft, curved shaft, branch-shaped and fork-shaped forgings. Due to the shape requirements, the billet-making processes of long shaft forgings include stretching, rolling, bending, pressing, forming, etc.

(1) The straight long shaft forging is a relatively simple type of forgings, which generally employs billet-making processes such as stretching, rolling (Fig. 4.11), pressing, and roll forging. The length of the billet obtained by billet-making process should be equal to the length of the final forging die cavity, and each cross-sectional area along the axis of the forging is equal to the sum of the cross-sectional area of the corresponding forging and the cross-sectional area of the flash. For a billet that can be simplified to a large diameter at one end and a small diameter at the other end, the upsetting or forward extrusion process can be used as billet-making process.

(2) For forgings with curved shaft, in addition to the above-mentioned billet-making steps, a bending step (Fig. 4.11) is required to make the shape of

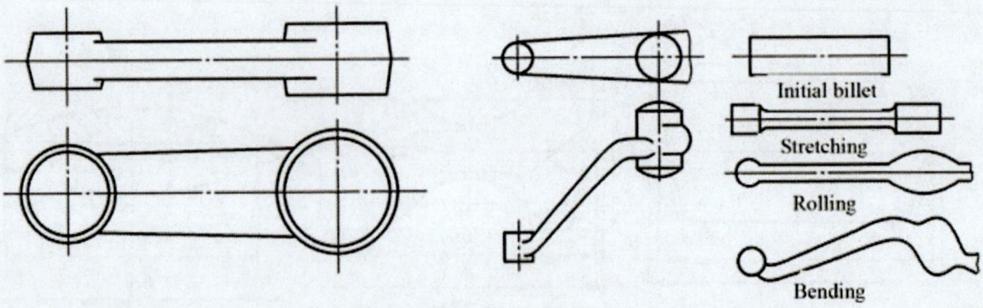

Fig. 4.11 Billet-making steps for straight long shaft and curved shaft forgings

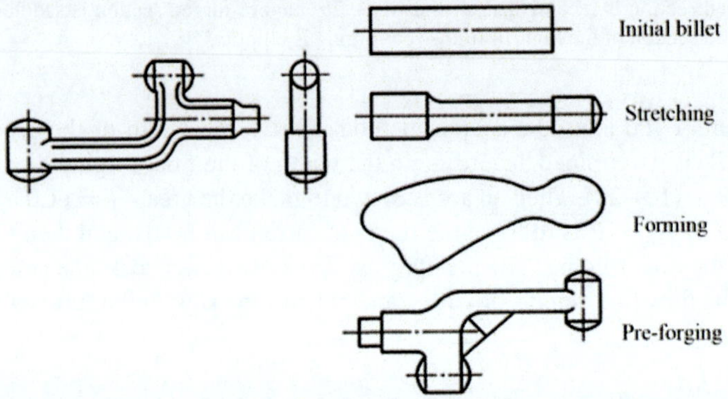

Fig. 4.12 Billet-making steps of long shaft forgings with branches

the obtained intermediate billet close to the shape of the final forging (or pre-forging).

(3) For the long shaft forgings with branches, the required billet-making steps are roughly the same as the previous ones, but a forming step (Fig. 4.12) or asymmetric rolling step is required to ensure that the branch cavity can be filled during pre-forging.

(4) For forgings with forks, in addition to the billet-making steps of straight long shaft forgings, the splitting step is generally required, as shown in Fig. 4.13. It can be seen that the billet-making steps of the long shaft forgings are determined according to the change of the axial cross-sectional area of the forgings, so that the metal distribution of the billet before the final forging is consistent with the requirements of the forgings.

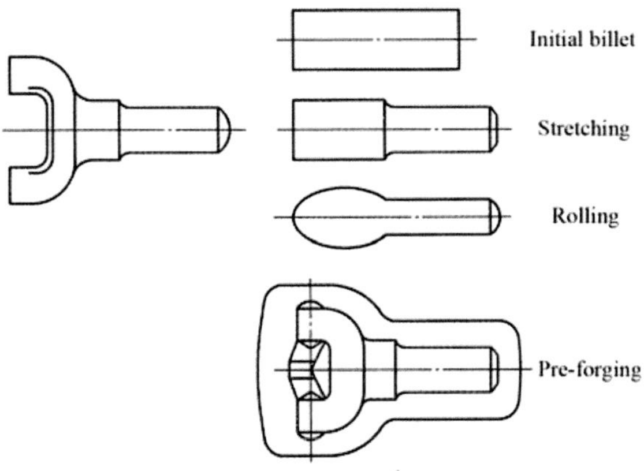

Fig. 4.13 Billet-making of the forging with forks

4.5.2 *Calculation of Billet Size*

The billet-making processes such as stretching, rolling, pressing, closed upsetting, and forward extrusion are usually designed based on the empirical calculation method.

The calculation of the billet is based on the assumption that the long shaft forging is in a plane strain state during forging. The length of the calculated billet is equal to the length of the forging. And the cross-sectional area in the axial direction A_c is equal to the sum of the cross-sectional area of the corresponding part of the forging A_p and the cross-sectional area of the flash A_f, that is

$$A_c = A_p + 2\eta A_f \tag{4.3}$$

where η is the filling factor, for simple forgings, it is 0.3–0.5, and for complex forgings, it is 0.5–0.8.

Generally, the billet size is calculated based on the cold forging. First, a number of representative cross sections is taken from the forging to calculate A_c, and then the cross-sectional change diagram is drawn, so the diameter of any part of the billet can be obtained from the following equation.

$$d_c = 1.13\sqrt{A_c} \tag{4.4}$$

After calculating each representative diameter, the diameter change diagram of the billet can be also obtained. The cross-sectional diagram and the diameter diagram are connected smoothly. The size of the billet can be determined according to the calculated results. The average cross-sectional area and average diameter of the billet can be calculated as follows:

$$A_{ave} = \frac{V_c}{L_p} \tag{4.5}$$

$$d_{ave} = 1.13\sqrt{A_{ave}} \quad (4.6)$$

where A_{ave} is the average cross-sectional area; d_{ave} is the average diameter.

4.6 Roll Forging

Roll forging are mainly used for aluminum alloys with better plasticity. Although the friction resistance of aluminum alloy is greater than that of ferrous metals, the forming processes are still similar [5, 6].

4.6.1 Principles and Characteristics of Roll Forging

Roll forging is a process developed on the basis of the rolling process (Fig. 4.14). During roll forging, parts are formed by continuous local deformation of the material. It must be pointed out that roll forging is different from conventional rolling, because the die cavity of the latter is directly engraved on the roll, and the sector modules of roll forging can be assembled and disassembled from the roll. In addition, the roll forged part is generally short.

The roll forging is a continuous quasi-static deformation process without shock and vibration. Compared with billet-making processes on a hammer, the productivity of roll forging process is 5–10 times that of hammer forging, the metal consumption of the material is reduced by 6–10%, and it is easy to realize mechanization and automation.

Because high-strength aluminum alloys have the characteristics of strong rate sensitivity, cracks are prone to appear when the billet is made on the hammer. Moreover, hot die forging presses and screw presses are not suitable for stretching. Therefore, roll forging is generally used to prepare billets for the forging process of long shaft aluminum alloy parts.

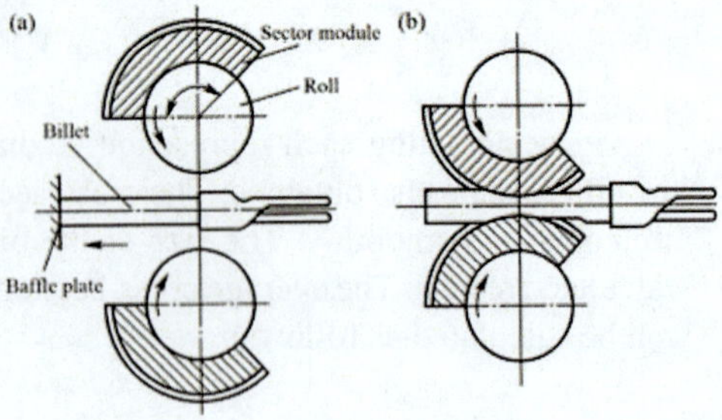

Fig. 4.14 Roll forging process: **a** initial state, **b** working state

4.6.2 Die Design of Roll Forging

The first in the design of roll forging process is to design the drawing of roll forged part. When designing, the drawing of roll forgings can be designed on the basis of the cross-section drawing of the forging. The following issues should be paid attention:

(1) According to the change of the cross-section area in the length direction of the forging, several representative section (Fig. 4.15b) are taken, such as the rod section and the head section of the connecting rod.

(2) In order to simplify the die structure and facilitate the calculation, the curves on the cross-section drawing are replaced with the corresponding straight line (Fig. 4.15c).

(3) The length of the end section of the roll forged part should be slightly shorter than the corresponding length of the forging, so that the part can be placed in the die forging cavity, and it can also avoid folding defects in the end section. The length of the middle section of the part should be the same as that of the forging.

The connecting region between the characteristic segments should be smoothly transitioned, otherwise it will produce folding defects in the roll forging and subsequent die forging. The transition region is generally included in the section with larger cross-section. The slope of the transition region is generally taken as 45°–60°. The length of the transition region is

$$l'_a = (0.5\text{–}0.86)(\sqrt{F} - \sqrt{F'}) \tag{4.7}$$

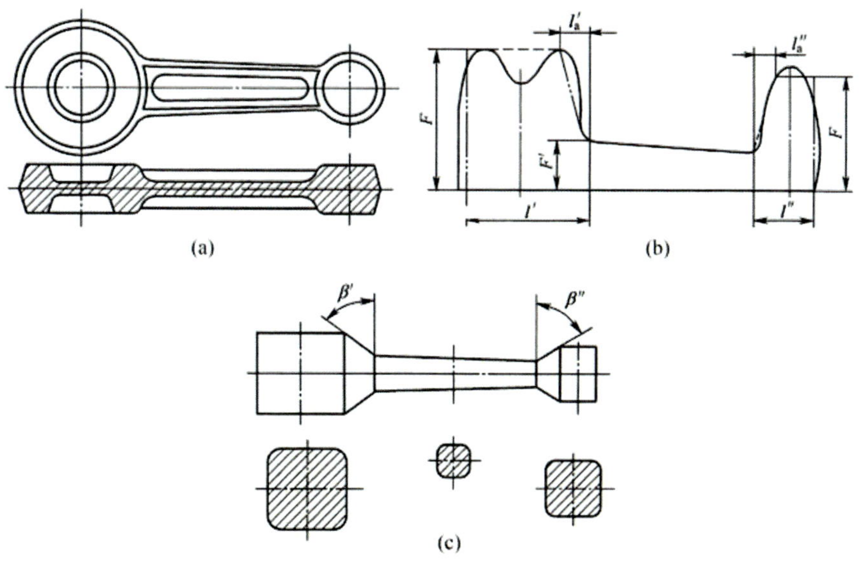

Fig. 4.15 Design of the roll forged part: **a** forging, **b** section drawing, **c** roll forged part

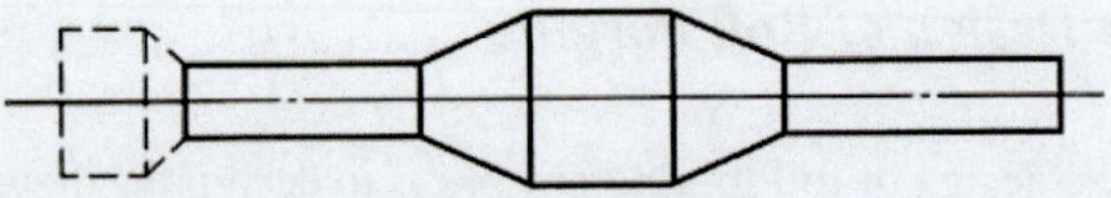

Fig. 4.16 Typical form of roll forged part (the dotted line indicates the clamp head)

where F and F' are the two characteristic cross-sectional areas of the transition region.

(4) The size of the roll forged part is selected according to the largest cross-section of the forging, which has two situations.

① The section with the largest cross-section is located at the end of the part. This section is not deformed during roll forging, and the section size is selected according to the size of the head.
According to the conditions of the roll forged part held by the clamps during roll forging, the length at the part end for clamping should not be less than 1/2 of the side length or diameter of the part. When the end of the part is not deformed, it can be used as a clamping section during roll forging, but its length must also meet the above clamping conditions.

② The non-deformed section is located in the middle of the part, as shown in Fig. 4.16. There may be two roll forging schemes: one is U-turn roll forging; the other is roll forging sequentially without the U-turn but needs a special clamp position at one end of the part. It can be seen that the material utilization of the first scheme is higher; and the production efficiency of the second scheme is higher.

(5) Cross-section drawing of roll forged part

In order to make the shape of die cavity not too complex, the cross-section of the roll forged part needs to be simplified. The principle of simplification is to change the curve to a straight line and keep the volume constant. According to the simplification, a section drawing of the roll forged part can be made, as shown in Fig. 4.17. It is the basis for die design and calculation.

The number of steps n indicates that it takes several rolls to reach the required shape and size. The total elongation coefficient λ_z is calculated as follows:

$$\lambda_z = \frac{A_0}{A_{min}} \tag{4.8}$$

where A_0 is the cross-section area of the original billet; A_{min} is the minimum cross-section of the roll forged part.

The total elongation coefficient is the product of the elongation coefficient for each roll forging step, namely

$$\lambda_z = \lambda_1 \lambda_2 \cdots \lambda_n \tag{4.9}$$

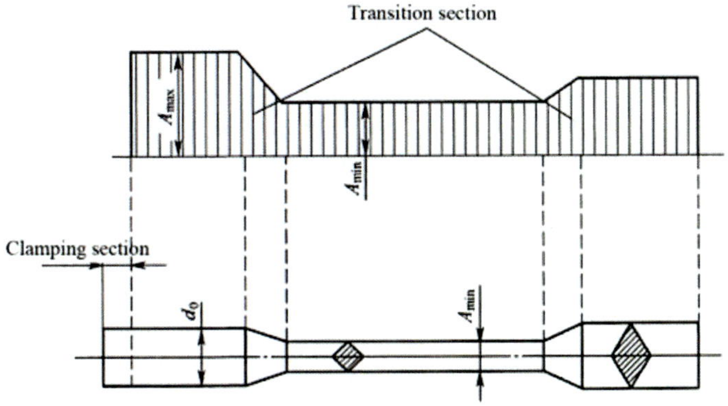

Fig. 4.17 Cross-section drawing of roll forged part

If the same elongation coefficient λ_{eq} is taken each step, then

$$\lambda_z = \lambda_{eq}^n \tag{4.10}$$

By taking the logarithm, the number of roll forging steps can be obtained.

$$n = \frac{\lg \lambda_z}{\lg \lambda_{eq}} \tag{4.11}$$

Generally, λ_{eq} takes 1.3–1.7.

The schemes of groove system available for roll forging are shown in Fig. 4.18.

The following principles should be considered when selecting the roll forging groove system.

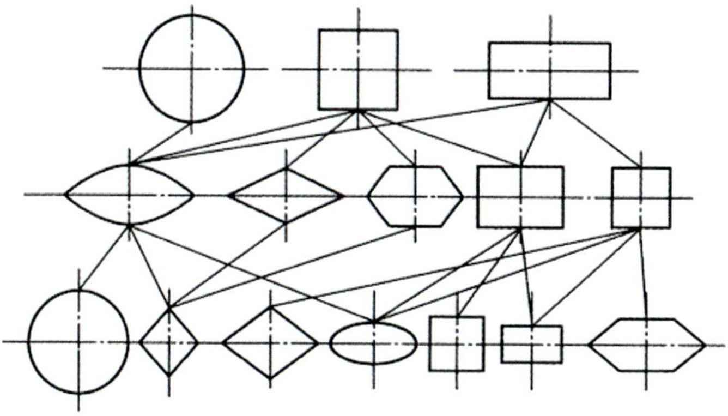

Fig. 4.18 Schemes of groove system for roll forging

(1) The cross-sectional geometry of the part obtained by roll forging should be conducive to the subsequent die forging. If the die forging requires the billet to have an elliptical cross-section, the "ellipse-square" or "ellipse-round" groove system must be considered during roll forging. Sometimes in order to make the shape of the billet meet the requirements of die forging, it is necessary to use a groove with a restricted width or a flash groove in the last step of roll forging.
(2) The selection of single-groove roll forging or multi-groove roll forging is determined by the elongation coefficient. In the multi-groove roll forging, it is often necessary to turn 90° or 45° when the part is moved to the next groove.
(3) The geometry of the original billet, such as square, round or rectangular, etc. should be considered.

4.6.3 Selection of Roll Forging Machine

The structure of roll forging machine is similar to that of a two-roll mill. It has a pair of forging rolls with the same rotation velocity and opposite directions. The typical structure of roll forging machine is shown in Fig. 4.19. The roll forging die is fixed on the forging roll. After the motor is decelerated by the transmission belt, the gear pairs drives the upper and lower forging rollers to rotate in the opposite direction at a constant velocity. Generally, there are friction clutches and brakes on the roll forging machine to obtain different operating specifications such as jog, single action, and linkage.

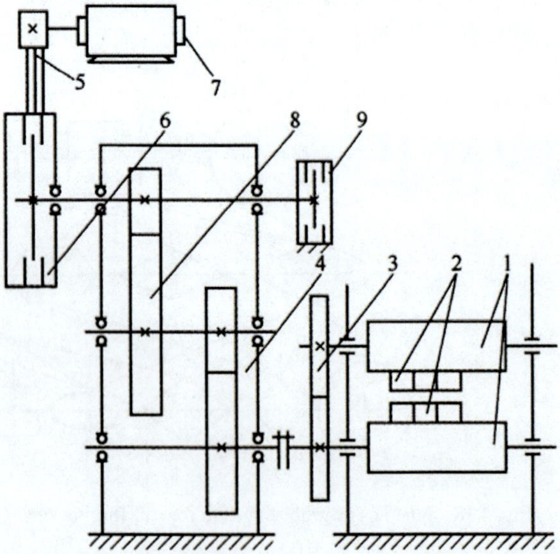

Fig. 4.19 Typical structure of roll forging machine. 1-forging rolls; 2-roll forging dies; 3, 4, 8-gear pairs; 5-transmission belt; 6-friction clutch; 7-motor; 9-brake

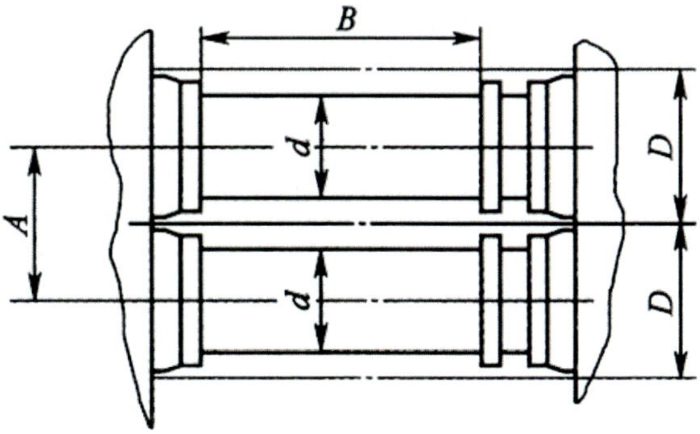

Fig. 4.20 Some technical parameters of the roll forging machine

According to the feeding direction, the structure of the forging roller, the relative position of the transmission part and the working part, the roll forging machine can be divided into cantilever type, dual-support type and compound type. In recent years, most of the roll forging machines are designed to automated roll forging units with manipulator operation. The cantilever type and dual-support type automated roll forging units are employed for long shaft aluminum alloy billets.

The technical parameters of the roll forging machine indicate its specifications, performance and main uses which are important for the selection of the roll forging machine. The main technical parameter of the roll forging machine is the nominal diameter D of the roll forging die, as shown in Fig. 4.20. Other technical parameters include nominal force P_g, forging roll diameter d, forging roll available length B, forging roll revolution n, forging roll center distance adjustment ΔA and forgeable side length of the billet H.

(1) Nominal diameter D of forging die refers to the nominal turning diameter at the parting surface of the forging die, and its value is equal to the nominal center distance A of the two forging rolls. The larger the nominal diameter of the forging die, the easier it is for the billet to be bitten. However, when the nominal diameter is increased, the size of the deformation zone increases, resulting in a significant increase in the roll forging force, which not only increases the size of the roll forging machine, but also increases the energy consumption during roll forging. Therefore, for certain roll forgings, the nominal diameter of the forging die must be selected reasonably so that it can not only meet the requirements of the roll forging process, but also has a reasonable equipment structure.

The nominal diameter of the forging die can be determined by analogy according to the size and shape of the forging, or it can be roughly calculated according to the billet diameter d_0.

During roll forging to prepare die forging billets, there are

$$D = (6\text{–}8)d_0 \tag{4.12}$$

During roll forging for part forming, there are

$$D = (8{-}15)d_0 \tag{4.13}$$

where d_0 is the diameter of the billet (mm).

(2) Nominal force p_g refers to the maximum radial load of the forging roll that the machine can withstand. It is compatible with the nominal diameter of the forging die.

(3) Forging roll diameter d refers to the diameter of the forging roll where the roll forging die is installed. The diameter of the forging roll determines the thickness of the forging die and the rigidity of the forging roll.

(4) Available length B for forging roll refers to the axial length of the part of the forging roll that can be installed with the dies excluding the clamping and fixing device at both ends. When the value is large, although the number of dies that can be installed is large, and the versatility is improved, the rigidity of the forging roll is reduced. Therefore, the available length of the forging roll should not be too large, usually $B = D$.

(5) The number of revolutions per minute of forging roll n should be coordinated with the number of strokes of its supporting die forging equipment. The rotating velocity of roll forging for forming should be adapted to its feeding device to ensure accurate and reliable feeding. The rotating velocity of the forging roll should be slightly lower than that of roll forging for billet-making.

(6) The adjustment amount ΔA of the center distance of the forging roll refers to the adjustment range of the center distance of the two forging rolls, and its value varies depending on the form of the machine adjusting mechanism. Generally ΔA takes 10–20 mm.

(7) Side length H of forgeable billet usually take the following

$$H = \frac{D}{6{-}8} \tag{4.14}$$

4.7 Die Design

4.7.1 Structural Design of Forging Die Used on Hot Die Forging Press

The hot die forging press is a static load equipment and the velocity is low, so multi-step die is often used to achieve the production of forgings. The multi-step die is composed of die units corresponding to each step and a general die base. Among them, the design of the die cavity and the die base is the key. Therefore, the following focuses on the design of the die cavity and the die base.

1. Key points of die cavity design

The most commonly used deformation steps on hot die forging presses are upsetting, bending, extrusion, pre-forging and final forging. Among these steps, the design of the pre-forging step is the most important. The shape and size of the pre-forging die cavity is quite different from that of the final forging die cavity, and the design has a great influence on the quality of the forging.

(1) Design of final forging die cavity

The design of the final forging die cavity is mainly to design the hot forging and the form and size of the flash groove. The design principles of hot forging and flash groove are basically the same as that of hammer die forging. However, because the die forging process on the hot die forging press employs a billet-making process, the deformation of the metal in the final forging die cavity is mainly carried out by upsetting, and the resistance of the flash is not as important as that of the hammer. The flash groove plays a more role in accommodating excess metal. Therefore, the height of the flash groove bridge part and the warehouse part is larger than that on the hammer.

When die forging is carried out on a hot die forging press, the height of the forging is ensured by the stroke of the press, and does not depend on the contact of the upper die and lower die on parting surfaces. Therefore, when the slider is at the bottom dead center, there must be a certain gap between the upper die and lower die to adjust the closing height of the die and offset the elastic deformation of the press to ensure the dimensional accuracy in the height direction of the forging. The gap between the upper die and lower die can also prevent the press from jam. The size of the gap depends on the size of the flash groove.

Since there is a gap between the upper and lower dies, and the die only bears the resistance of metal plastic deformation. Therefore, a small insert forging die with higher hardness can be used to replace the overall forging die.

(2) Design of pre-forging die cavity

The pre-forging die cavity is designed according to the pre-forging. The general principle of designing the pre-forged part is to make the part upset as much as possible in the cavity of the final forging die. The cross-sectional area of the pre-forged part is slightly larger than the corresponding cross-sectional area of the final forged part. The height of the pre-forging part is 2–5 mm larger than that of the final forged part, and the width is reduced accordingly. If the cross-section of the final forged part is circular, the cross-section of the corresponding pre-forged part is also circular, and the roundness of the cross-section is 4–5% of the diameter of the cross-section of the final forged part. The positioning of the pre-forged part in the final forging die cavity should be considered. For this reason, the shape and size of certain parts in the pre-forging should be basically consistent with the final forging.

2. Design of die base

The design of the die base mainly includes the design of the following components: upper template and lower template, upper backing plate and lower backing plate, upper module and lower module, ejector devices, guide devices, and some fasteners.

The structural form of the die base should have good versatility to adapt to the production of multiple varieties. The die base should have sufficient strength and rigidity so that the elastic deformation caused by the equipment during the forging process will not affect the dimensions of the forging in the height direction. Therefore, the parts in the die base that bear the forging load should be made of alloy steel and undergoes reasonable heat treatment. The shapes of all parts in the die base should be as simple as possible to facilitate manufacturing. The ejector devices in the die base should be reliable, durable, and easy to repair and replace. The structure of the die base should ensure that it does not need to be removed from the press when installing, adjusting and replacing modules to save time. All the fasteners in the die base should be arranged properly and easy to operate. Lifting holes or lifting rods should be provided on the die base.

Figure 4.21 shows a typical die base used to produce long shaft forgings on a hot die forging press. There are three sets of ejector rods in the pre-forging die and the final forging die, respectively. Therefore, two three-arm ejector levers are installed in the upper and lower templates of the die base, and four separable bearing seats are added correspondingly.

The typical structure of forging die used on the hot forging press is shown in Fig. 4.22. The dimensions in the height direction should generally conform to the following proportional relations:

(1) $H = A + 0.75a$, where H is the closed height of the die; A is the minimum closed height of the press; a is the adjustment amount of the closed height of the press.
Taking into account that the repair of the die base will reduce the height, and in order to compensate for the elastic deformation of the press and certain clearances, 75% of the adjustment is recommended.
(2) $h_1 + h_2 = (0.3 - 0.325)H$, where h_1 is the thickness of the template of the die base; h_2 is the thickness of the backing plate.
The thickness of the template h_1, backing plate h_2 and module h_3 are related to each other and determined according to the strength conditions within a certain height range of closed die. When there is no change in h_2, the increase of h_1 will reduce h_3. On the contrary, the increase of h_3 will weaken the strength of the entire die base. Therefore, it is necessary for these variables to have an appropriate values.
(3) The gap between the upper module and lower module is $h_n = h_f$.
(4) The closed height of the die $H = 2(h_1 + h_2 + h_3) + h_n$.

Modules are available in cylindrical and rectangular (or square) forms. Cylindrical modules are easy to manufacture, and are suitable for revolving forgings. Rectangular modules are suitable for forgings of any shape. The size of the module depends on

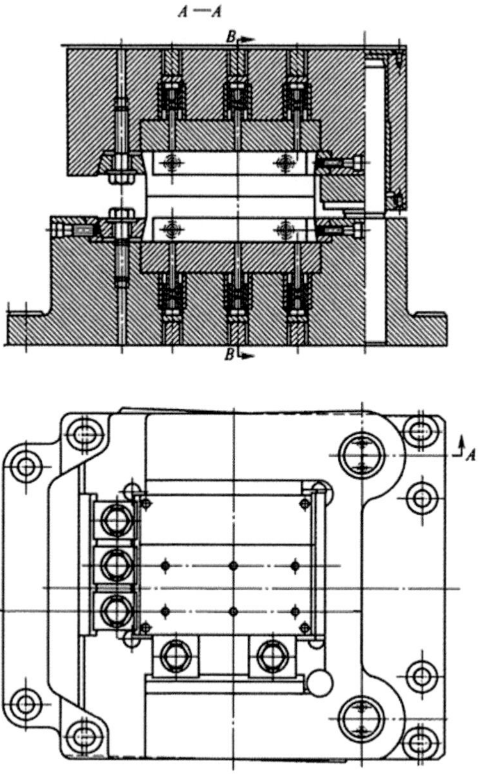

Fig. 4.21 Typical die base used for long shaft forgings

Fig. 4.22 Typical forging die used on the hot die forging press. 1-upper template; 2, 8-backing plate; 3, 6, 9, 11-fastening plate; 4-upper module; 5-lower module; 7-lower template; 10-guide device

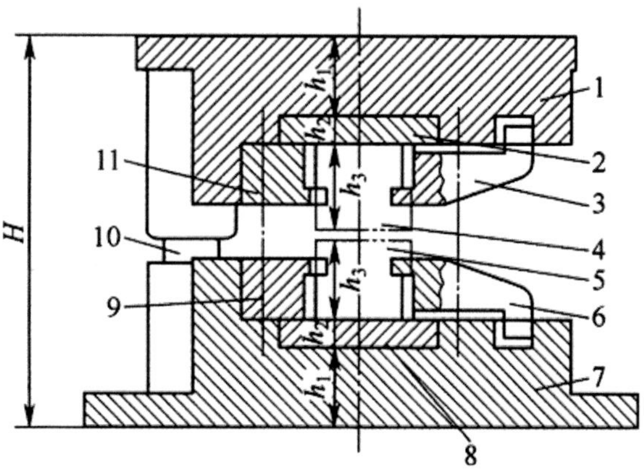

the size of the die cavity and the thickness of the die wall. The thickness of the die wall t can be determined as follows:

$$t = (1{-}1.5h) \geq 40\,\text{mm}$$

where h is the depth of the widest part of the cavity.

The thickness s between the bottom surface of the module and the deepest part of the die cavity should not be less than $(0.6{-}0.65)h$. There should be a margin for renovation when determining the height of the module, but the total height should not be greater than $(0.35{-}0.4)H$.

If there is a deep cavity in the module, a vent hole should be set at the place where the deep cavity is filled last. The hole diameter is 1.2–2 mm, the hole depth is 20–30 mm, and it is connected with the hole with the diameter of 8–20 mm until the bottom of the module. When there is an ejector or other venting gap at the bottom of the die cavity, there is no need to set a vent hole.

3. Ejector device

Generally, there is an ejector in the modules, which is used to eject the forging from the die cavity. The configuration of the ejector should be determined according to the shape and size of the forging. The position of the ejector on three different shapes of forgings is shown in Fig. 4.23. The ejector shown in Fig. 4.23a acts on the flash. The ejector shown in Fig. 4.23b acts on the punching wad with a large diameter. And the ejector shown in Fig. 4.23c acts on the small end and the big end of the forging, respectively.

The diameter of the ejector should not be too small, otherwise it is easy to bend, generally ϕ 10–30 mm. There should be a guide part of sufficient length on the module with a gap of 0.1–0.3 mm between the ejector rod.

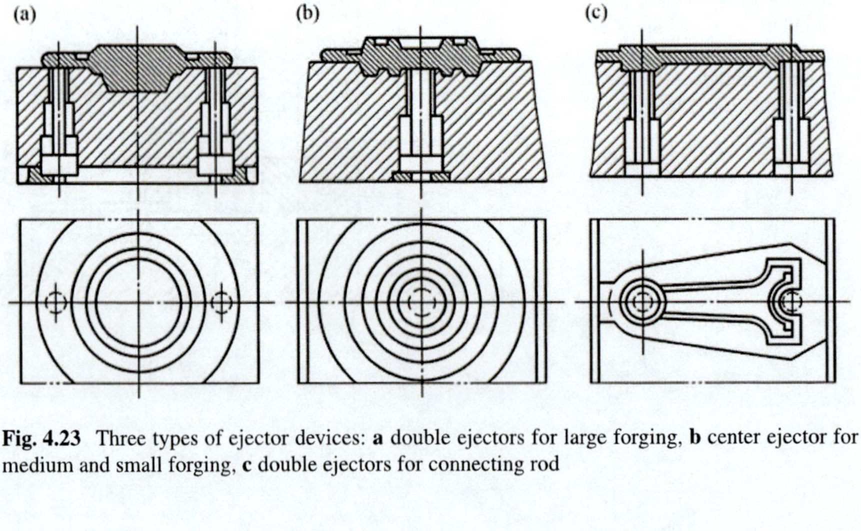

Fig. 4.23 Three types of ejector devices: **a** double ejectors for large forging, **b** center ejector for medium and small forging, **c** double ejectors for connecting rod

4. Guiding device

The guide devices of forging die is composed of guide posts, guide sleeves and other parts. Most forging dies use double guide posts which are located on the back side of the die base. The guide posts and the guide sleeves are transitionally matched with the upper die base and lower die base, respectively. The gap between the guide sleeve and the guide post is 0.25–0.50 mm.

4.7.2 *Structural Design of Forging Die Used on Screw Press*

1. The structure of forging die

The impact characteristics of the screw press are generally between the die forging hammer and the hot die forging press. Therefore, the die structure can be integral, as shown in Fig. 4.24a and b, or combined, as shown in Fig. 4.24c and d. Because the combined forging die can save die steel, facilitate standardization of die parts, and reduce costs, it is widely used for the mass production of small and medium forgings. However, the integral die structure is often used on large-tonnage screw presses.

2. Design of die cavity and module

When there is only one die cavity on the module, the center of the die cavity and the center of the module coincide with the center of the main screw of the screw press. When there are two cavities for pre-forging and final forging on the module, the center of the final forging die and the center of the pre-forging die should be placed on both sides of the center of the module. The distance between the two centers

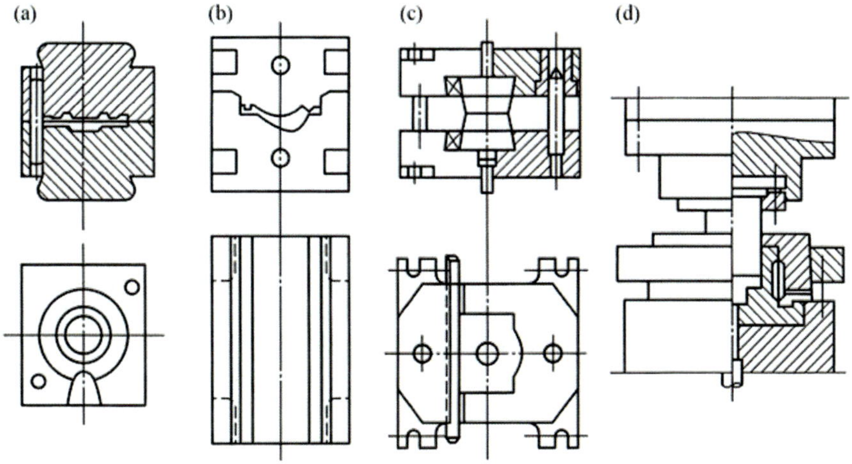

Fig. 4.24 Common forging die structure used on screw press: **a**, **b** integral type; **c**, **d** combined type

relative to the center of the module is $a/b \leq \frac{1}{2}$, and $a+b \leq \frac{D}{2}$, where D is the screw diameter of the screw press. When there are two final forging die cavities at the same time, it should be kept $\frac{a}{b} = 1$, and $a + b \leq \frac{D}{2}$.

Because the velocity of the screw press is slower than that of the die forging hammer, and the stress conditions of die are better, the bearing area of modules can generally be 1/3 of that on the hammer. Since screw presses are generally equipped with lower ejector devices, the cavity with a more complicated shape is set in the lower die. Modern large-tonnage electric screw presses are also equipped with ejector devices on the slider. When designing the die cavity and the module, the structure should be simple, and the production cycle is short to achieve the best economic effect under the condition of ensuring the strength. For the parts of which the die cavity is relatively deep and complex, vent holes should be provided.

If the billet-making process is used to make the metal distribution reasonable and the small flash die forging is used, the flash groove shown in Fig. 4.4 can be used. For some small forgings, the flash groove form of Fig. 4.2a can be used. For complex forgings, the flash groove form of Fig. 4.2c can be used.

For safe production, it is very important to correctly design the fastening form of modules. The modules can be fasten by wedge, pressure ring and bolt. The fastening form of wedge is convenient and reliable, and is generally used for larger modules. The fastening form of pressing ring is only suitable for round modules, and it is also convenient and reliable. In the fastening form of bolt, the module can be round or rectangular. Its advantages are simple structure and convenient manufacturing. But it should be noted that the bolts are easy to loosen during the forging process.

From the perspective of minimizing the misalignment of the forging die, and improving the service life of the main screw and guide rail, the use of single-die forging is the most reasonable. However, in order to extend the scope of application of the screw press, according to the process characteristics of the die forging and to improve the utilization of the equipment, a double-cavity is often used. Modern large-tonnage electric screw presses have been developed to the three-step die forging, that is, there are three cavities.

In order to balance the misalignment force during the forging process and improve the forging accuracy, a guide device can be used. There are four guiding types for forging die on screw press: guide lock, guide pin, guide pin and guide sleeve, and punch and die guides.

During the design of closed forging die without flash, the punch and the die, as well as the ejector pin and the die need to be able to slide freely. Therefore, there must be an appropriate gap. If the gap is too large, it will generate longitudinal flash which accelerate the wear of the die and cause the difficulty of ejection. If the gap is too small, the deformation of the die will make the movement difficult. Usually the gap between the ejector pin and the die is selected according to the accuracy of the clearance fit. The clearance between the punch and the die is listed in Table 4.4.

Table 4.4 Clearance between the punch and the die (mm)

Punch diameter	Clearance
<20	0.10
20–40	0.10–0.15
40–60	0.15–0.20
>60	0.20–0.30

4.7.3 Design of Trimming Die and Punching Die

Whether it is the traditional open die forging or the small flash precision forging, there is a circle of flash around the parting surface of forgings, sometimes as well as a punching wad. After die forging, it is necessary to use a trimming die and a punching die to cut off the flash and the wad. Hot trimming and hot punching are prone to longitudinal burrs, so cold trimming and cold punching are usually used for aluminum alloy forgings.

As shown in Fig. 4.25, the trimming die and punching die are mainly composed of a punch and a die. When trimming, the forging is placed on the die, and the flash of the forging is cut by the cutting edge of the die and separated from the forging under the push of the punch. When punching, the situation is the opposite. The punching die only plays a supporting role, while the punch plays a shearing role.

Trimming die and punching die are divided into three types: simple die, progressive die and compound die. The simple die is used to realize trimming or punching in one stroke of the press. The progressive die is to simultaneously realize trimming of one forging and punching of the other forging in one stroke of the press. The compound die is to realize trimming and punching of one forging in one stroke of the press.

The choice of structure type is mainly based on factors such as production, and trimming and punching methods. For small batch production, the simple die should be used; for mass production, the progressive die or compound die should be used.

1. Trimming Die

The trimming die is generally composed of trimming cavity, trimming punch, die base, stripper device and other parts.

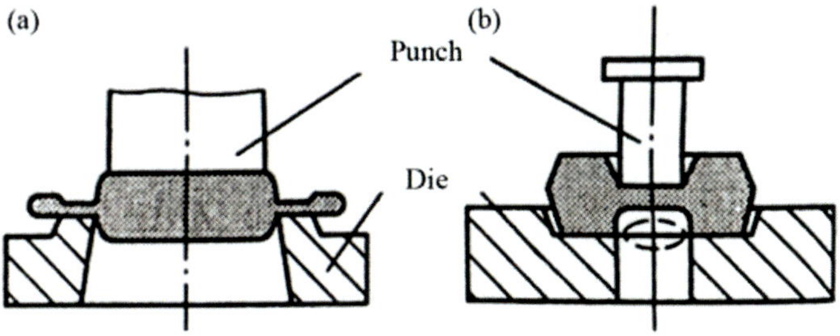

Fig. 4.25 Schematic of trimming die (**a**) and punching die (**b**)

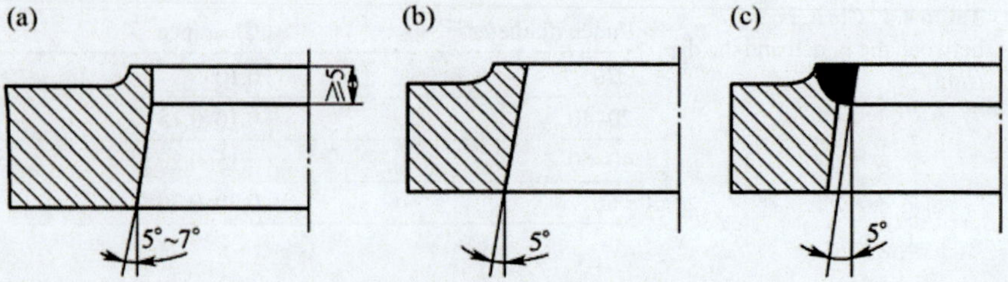

Fig. 4.26 Cutting edges: **a** straight edge; **b** oblique edge; **c** surfacing edge

The trimming die has two structures: an integral type and a combined type. The former is suitable for medium and small forgings, especially forgings with simple and symmetrical shapes. The later is composed of more than two cavity dies, which is relatively easy to manufacture and adjust. It is mostly used for large or complex forgings.

There are three types of cutting edges. Figure 4.26a shows a straight cutting edge. When the cutting edge is worn, the top surface can be ground off to restore the sharpness, and the size of the edge contour remains unchanged. Although the straight edge is convenient to maintain, it has a large cutting edge force and is generally used for integral dies. Figure 4.26b shows an oblique cutting edge which can be processed by a slotting machine. The trimming of this cutting edge is labor-saving, but it is easy to wear. It is mainly used for combined die. The joint surface of the divided die can be ground off or repaired by surfacing after the cutting edge is worn. Figure 4.26c shows another cavity die composed of a base body and a cutting edge. The base body is cast with cast steel and the cutting edge is surfacing with die steel, which can greatly reduce the cost of the die.

In order to facilitate the placement of forgings, the top surface of the cutting edge should be made into a boss form. The dimensions of the trimming die are listed in Table 4.5, where $B_{\min}$ is the minimum wall thickness, $H_{\min}$ is the minimum height of the die, and H is the edge height of the die. The width of the boss is equal to the width of the flash groove bridge b or $b-(1-2)$ mm.

The trimming die is usually fastened to the die base with wedges or screws. The wedge is simple and mostly used for an integral die or a two-piece die. The screw fastening method is mostly used for combined dies of more than two pieces, which is convenient for adjusting the position of the cutting edge.

Table 4.5 Dimensions of trimming die

Height of flash bridge	$H_{\min}$	H	$B_{\min}$	Remarks
<1.5	50	10	30	1000 kN trimming press
1.5–2.5	55	12	35	3150 kN trimming press
>3	60	15	40	3150 kN trimming press

Table 4.6 Clearance between the punch and the die for trimming

Clearance between the punch and the die when the forging slope is smaller than 15°		Clearance between the punch and the die of circular cross-section forgings	
h	δ	*D*	δ
<10	0.5	<30	0.5
10–18	0.8	30–47	0.8
18–24	1.0	47–58	1.0
24–30	1.2	58–70	1.2
>30	1.5	>70	1.5

The trimming punch plays the role of transmitting pressure, so it needs to have a certain contact area with the forging. Uneven contact surface or too small contact surface will cause defects such as bending, distortion and surface crushing.

The clearance between the punch and the die δ is determined according to the cross-sectional shape and size of the forging perpendicular to the parting surface and listed in Table 4.6, where h is the height of the forging, D is the diameter of the forging. If the clearance is too large, it is not conducive to the alignment of the punch and the die. Moreover, it is easy to generate eccentric trimming and uneven residual burrs. If the clearance is too small, the flash is not easy to be removed from the punch, and the punch and the die may gnaw each other. When the forging slope is greater than 15°, the clearance should not be too large, so as not to cause the edge of the forging to roll up.

If the clearance between the punch and the die is small, and the punch needs to enter the die during trimming, the flash is often stuck on the punch after trimming and is not easy to strip. Therefore, when the clearance of cold trimming is less than 0.5 mm and the clearance of hot trimming is less than 1 mm, a stripper device is required. There are two types of stripper devices: rigid stripper device and elastic stripper device. Figure 4.27a shows a commonly used structure which is suitable for the trimming of small and medium forgings. Figure 4.27b shows a claw-shaped stripper device which is suitable for the trimming of large and medium forgings. For

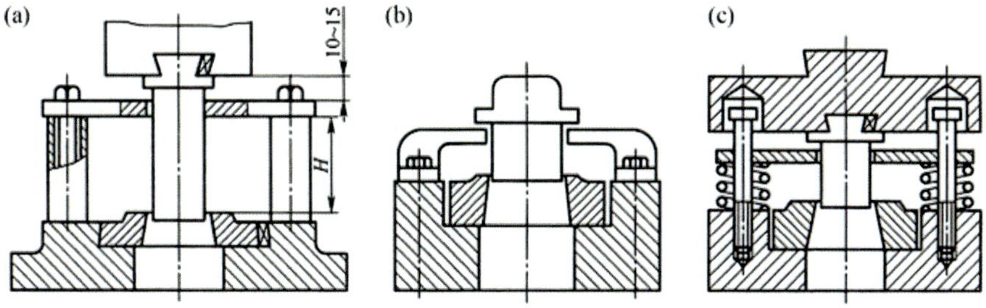

Fig. 4.27 Stripper devices: **a** rigid stripper device; **b** double claw-shaped rigid stripper device; **c** Elastic stripper device

forgings with larger height, if the shoulder of the punch will hit the stripper plate after the die is closed, the elastic stripper device shown in Fig. 4.27c can be used.

The distance between the stripper plate and the cutting edge should be able to ensure that the forging can be placed freely. When the die is closed, there should be a distance of 10–15 mm between the shoulder of the punch and the stripper plate. The hole size of the stripper plate is designed to increase by 1 mm on each side based on the size of the punch. The thickness of the stripper plate is generally 10–20 mm.

2. Punching die

When punching the wad, the forging can be placed in the punching die, and the wad is cut by the edge of the punch. The size of the cutting edge of the punch is determined by the hole size of the forging. The die plays the role of supporting the forging, and the forging is positioned through the cavity. The size of the cavity in the vertical direction is determined by the size of the corresponding part on the forging, and its maximum depth does not need to exceed the height of the forging. For forgings with symmetrical shapes, the depth of the cavity can be slightly smaller than half of the corresponding height of the forging. When designing the size of the cavity in the horizontal direction, there should be a clearance Δ between the side of the positioning part (dimension C in Fig. 4.28) and the forging, with a value of $\frac{e}{2} + (0.3\text{–}0.5)$ mm, where e is the positive deviation of the forging. In the non-positioning part (dimension B in Fig. 4.28), the clearance Δ_1 can be larger than Δ, $\Delta_1 = \Delta + 0.5$. The bottom surface of the forging should be fully supported on the die, so the diameter of the die d should be slightly smaller than the diameter of the inner hole on the bottom of the forging. The minimum height of the die hole H_{min} should not be less than $s + 15$, where s is the thickness of the wad.

During the design of the trimming and punching compound die (Fig. 4.29), the clearance between the punch and the die δ_1 is not less than 1 mm, and the punching

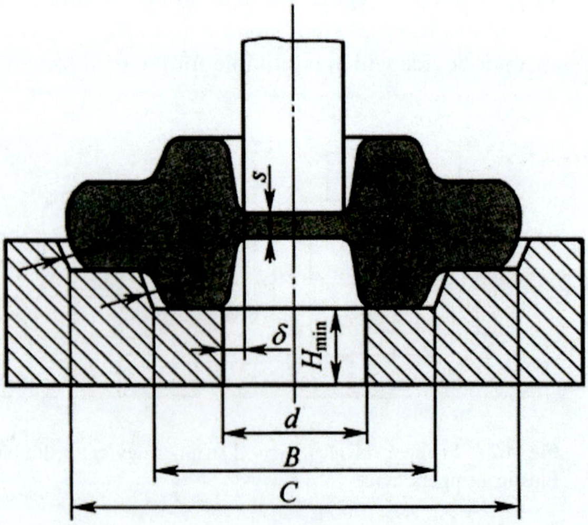

Fig. 4.28 Dimensions of the punching die

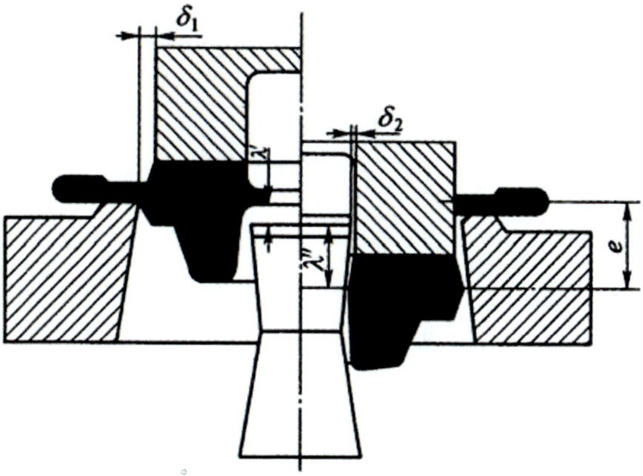

Fig. 4.29 The dimensions of the trimming and punching compound die

clearance δ_2 is not less than 0.6 mm. If the clearance is too small, a device for strip flash and wad should be designed. The trimming process should be completed before punching to reduce working pressure. There should be a proper clearance between the wad and the punch $\lambda' = 5-15$. The cutting stroke e should be able to ensure that the flash and wad are fully cut, and the distance λ'' is usually 10–15 mm.

4.7.4 Design of Sizing Die

Aluminum alloy forgings are prone to deformation such as bending and torsion during the process of trimming and punching. Therefore, it is necessary to correct the dimension of the forgings. The forgings that need to be corrected include slender shaft forgings that are prone to bending, long rod forgings that are easy to bend or twist at both ends in the vertical direction of the parting surface, fork-shaped and branch-shaped forgings that are prone to deformation, slender forgings with curved parting surfaces, forgings with large dimension change, forgings with thin flanges, forgings with complex shapes, such as shift forks, crankshafts and camshafts.

The correction of forgings can be hot sizing or cold sizing. The hot sizing is usually carried out after the hot trimming and punching. It is generally used for large forgings. Cold sizing is used as the final process, generally used for small and medium forgings.

The cavity of sizing die is designed according to the forging (hot or cold). In order to make it easier to put the forging in or take out of the die, and consider that the lateral size of the forging will increase during the sizing process, there should be a certain clearance Δ_1 between the die and the forging in the horizontal direction, and its value is related with the cross-sectional shape and size of the forging, which can be selected from Table 4.7. For forgings with high boss ($H/D > 1$), the clearance can be large. For easily deformable fork-shaped forgings, in order to obtain a good sizing effect, no clearance is set in the top part of the fork. For small forgings, the height

Table 4.7 The clearance between the die cavity and the forging (mm)

Section shape	Clearance Δ_1					
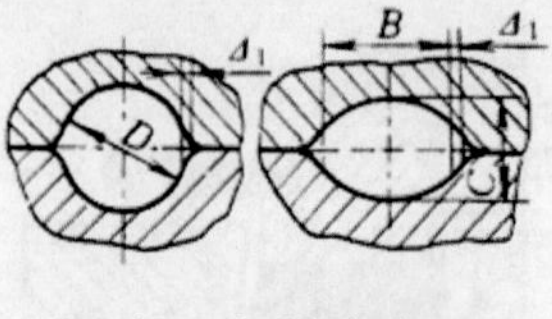 Circular and elliptical cross-sections	*D* or *B*	≤10	11–20	21–40	41–60	>60
	Δ_1	0.8	1.0	1.5	2.0	2.5
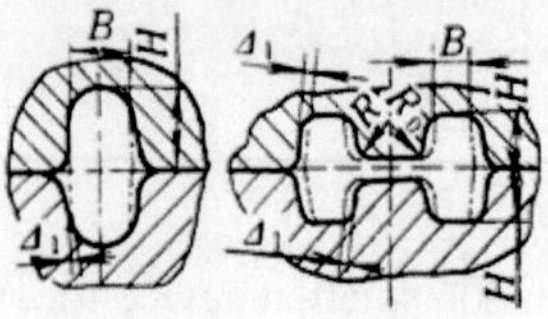 I-shape and rib-web section	*H*	≤10	11–20	21–30	31–40	
	Δ_1	1.2	1.5	2.0	2.5	
	$R = R_0 + (2\text{–}5)$					
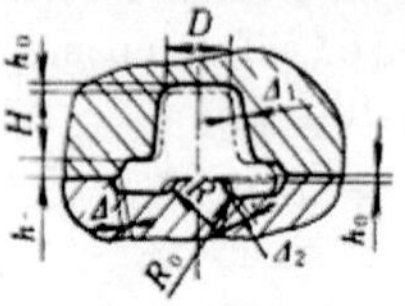 General shape	*H*	≤30	31–45	46–60	>60	
	Δ_1	1.5	2.0	2.5	3.0	
	Δ_2	0.8	1.0	1.0	1.0	
	When $H/D > 1$, the above data are employed. When $h_0 =$ 1–4 mm, the *R* is the same as that of I-shape and rib-web section					

of the die cavity is equal to the height of the forging. For large and medium-sized forgings, the height of the die cavity is smaller than the height of the forging, and the difference is the negative deviation of the forging. The edge of the die cavity should be rounded with a radius of 3–5 mm. For forgings with complex shapes, such as crankshafts, camshafts, etc., two cavities must be used for sizing in two directions. The structure of the sizing die is similar to that of the final forging die. In addition, there is a gap of 1–2 mm between the upper and lower die.

4.8 Examples of the Precision Forging Process for Typical Parts

As mentioned above, there are many varieties of long shaft aluminum alloy forgings with different shapes. This book presents a few representative examples to introduce the forging process, key technologies and process verification.

4.8.1 *Multi-step Precision Forging of 2014 Aluminum Alloy Connecting Rod*

The connecting rod has a flat connecting rod structure, as shown in Fig. 4.30.

1. Die forging process

According to the plasticity diagram of 2014 aluminum alloy, the process performance of this alloy is better in the temperature range of 300–450 °C, and the plasticity on a hot die forging press is higher than that on a hammer. Its deformation resistance increases with the decrease of the deformation temperature and the increase of the strain rate, which is the same as other aluminum alloys. In addition, it can be seen from the recrystallization graph that the alloy can be deformed within a relatively large deformation degree. The designed forging process consists of cutting, heating, billet-making, second heating, die forging, cold trimming and punching, cold sizing, acid wash, milling, heat treatment and quality inspection.

The bar was cut with band sawing machine. When it was processed by cross wedge rolling to make billets, the bar specification was ϕ 40 × 80. When it was processed by closed upsetting, the bar specification was ϕ 30 × 150. Box type resistance furnace or mesh-belt type resistance furnace with forced air circulation device and temperature control system was used for heating. The heating temperature was 450 °C, and the holding time was 60 min. The billet could be made by cross wedge rolling or closed upsetting, which was selected according to the size of the production batch. When the billet was prepared by cross wedge rolling, one cut bar could be rolled into two billets, the billet that enters the subsequent die forging first could be directly forged, and the other billet needed to be heated again. The heating temperature was 450 °C, and the holding time was 35–40 min.

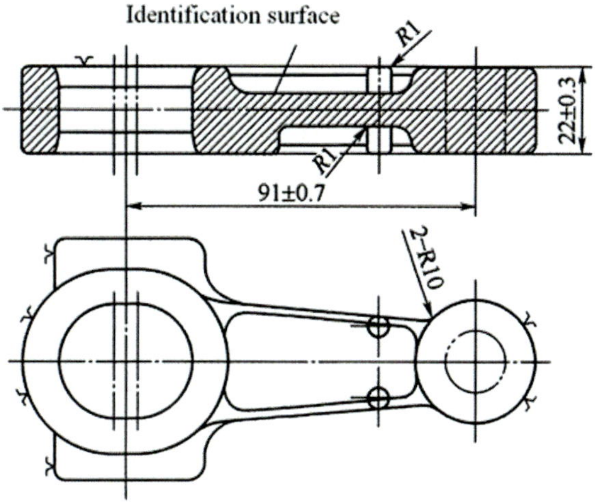

Fig. 4.30 The forging drawing of 2014 aluminum alloy connecting rod

The forging process included pre-forging and final forging. For connecting rods with large internal slope and transition fillet, the larger diameter end of the billet was first squashed to a height of 20 mm, and then the billet was deformed in the final forging die. For connecting rods with small internal slope and transition fillet and large ratio of rib height to thickness, the process included flattening, pre-forging and final forging.

After final forging, a 600–1000 kN press and a trimming-punching compound die were used for trimming and punching. Then, a 4000 kN screw press was used for cold sizing to eliminate the deformation caused by trimming and punching.

The purpose of acid wash was to remove residual lubricant on the surface and make the surface bright. When longitudinal burrs appeared on the cutting edge, a fine-grained grinding wheel was used to polish the burrs. Finally, the forgings were heat treated, and the hardness after T6 treatment was 110–130 HB.

2. Analysis of die forging process

There were two die forging schemes for 2014 aluminum alloy connecting rods. One was to use cross wedge rolling to prepare the billet and then small flash precision forging to form the part. The other was to use a closed upsetting to prepare a billet, and then pre-forging and final forging.

(1) Scheme 1

Figure 4.31 shows the billet drawing obtained by the cross wedge rolling of the bar of ϕ 40 × 158. Because the volume distribution of the billet produced by the cross wedge rolling was in good agreement with the cross-sectional change along the length of the connecting rod, the large end and small end of the rolled billet can be flattened to a height of about 20 mm, and then the billet was final forged. The flash groove of the final forging die cavity was the small flash groove shown in Fig. 4.4. When the small flash was employed, the lateral resistance increases rapidly, which can force the longitudinal flow to increase. Therefore, for this kind of small flat connecting rod, the pre-forging step was not be used.

The advantage of this scheme was that the material utilization can be increased to more than 88%. The disadvantage was that the equipment cost is relatively large due

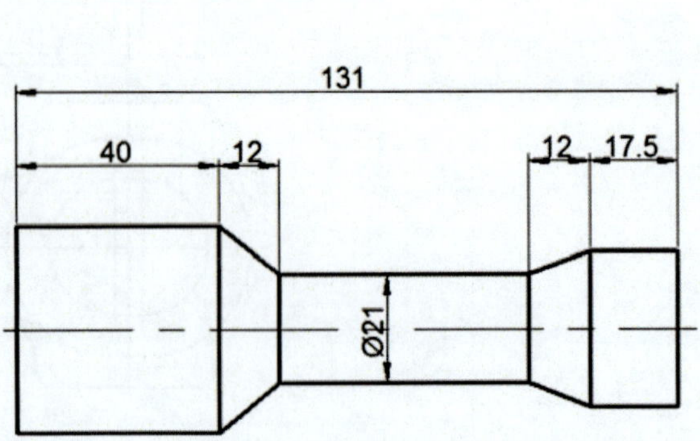

Fig. 4.31 Billet drawing of cross wedge rolling for the connection rod

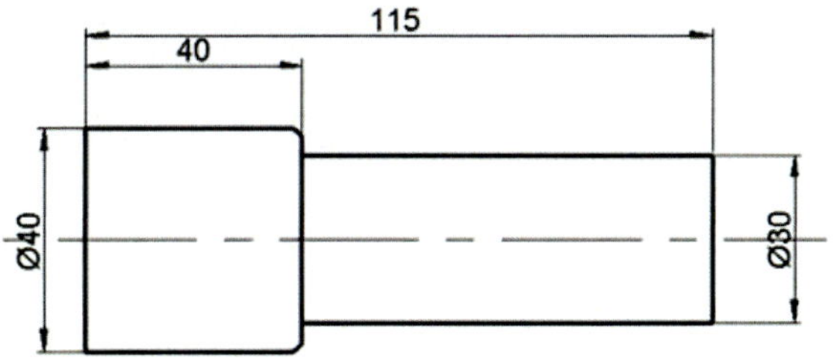

Fig. 4.32 Billet drawing of closed upsetting for the connection rod

to special cross wedge rolling equipment. In addition, the number of heating might increase. Therefore, this scheme was suitable for mass production.

(2) Scheme 2

Figure 4.32 shows a billet drawing obtained by closed upsetting of a bar of ϕ 30 × 150. It was obtained by simplifying the calculated billet shape to the head-rod shape under the condition of constant volume. Since the upsetting ratio of the head was 2.5 greater than the allowable upsetting ratio of the screw press, closed upsetting could be employed.

The rod part of billet was simplified according to the calculated billet diameter. Its volume distribution was quite different from the actual size of the connecting rod. Therefore, a pre-forging step was required. In addition, because there was more excess metal in the rod, the traditional die forging process with conventional flash form was used.

The pre-forging and final forging process were simulated by a thermal–mechanical coupling finite element software DEFORM-3D. The simulation results of the pre-forging is shown in Fig. 4.33. The pre-forging results was good, and the effective stress, effective strain and temperature field were within a reasonable range, which indicated that the closed upsetting was reasonable and feasible for billet-making.

The simulation results of the final forging is shown in Fig. 4.34. The forging was well formed, and the effective stress, effective strain and temperature field were also within a reasonable range, indicating that the process plan was reasonable and feasible.

Process test was carried out according to scheme 2. The obtained forging is shown in Fig. 4.35. The quality inspection indicated that there are no internal defects in the forgings, and the dimensional accuracy, surface quality, hardness and mechanical properties of the forgings have all reached the specified values.

Although the material utilization of scheme 2 is 75% which was lower than that of scheme 1, its total manufacturing cost was lower than that of scheme 1. Therefore, when the batch of forgings is not very large, the scheme 2 has better application value.

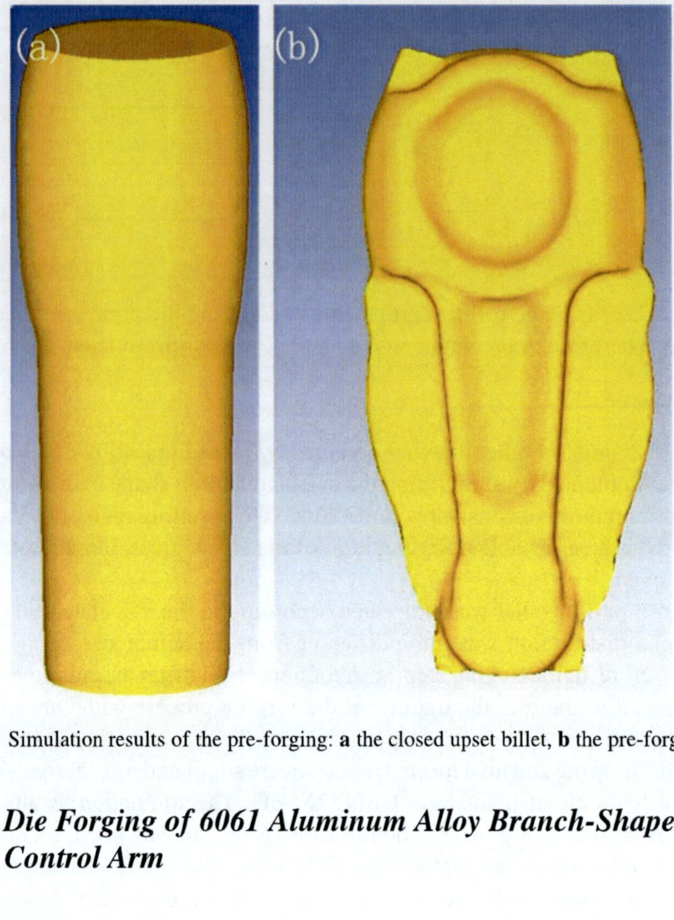

Fig. 4.33 Simulation results of the pre-forging: **a** the closed upset billet, **b** the pre-forged part

4.8.2 Die Forging of 6061 Aluminum Alloy Branch-Shaped Control Arm

1. Die forging process

The control arm is the main structural and force-bearing component in the automobile suspension system. Its shape is extremely complex with the comprehensive structural characteristics of long shaft, disc-shaped and branch-shaped. The control arm not only has to support the weight of the car body, but also has to withstand the steering torque and the braking torque when braking [7].

A typical aluminum alloy control arm is shown in Fig. 4.36. It can be seen that the cross-sectional area of the forgings along the axis changes obviously. The basic distribution law is that the middle cross-sectional area is large, and the cross-sectional area at both ends is small. The forging is a three-dimensional curved shape, therefore, a curved parting surface was used instead of a flat parting surface. There is a branch and a high boss in the middle of the forging where a lot of material is needed.

Because the shape of the control arm is complex, the pre-forging process was required to ensure accurate forming during final forging. The forging process route

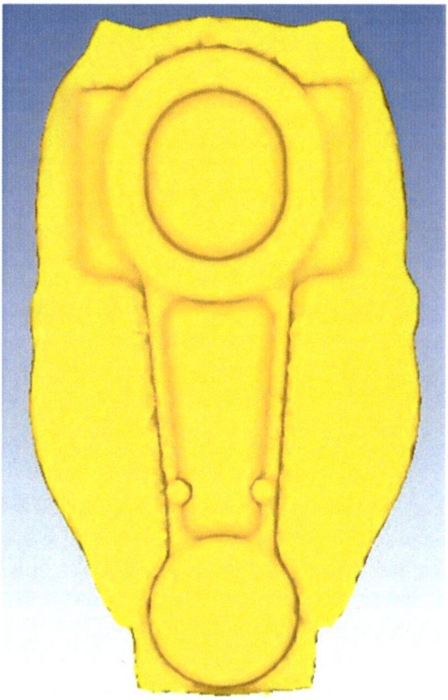

Fig. 4.34 Simulation results of the final forging

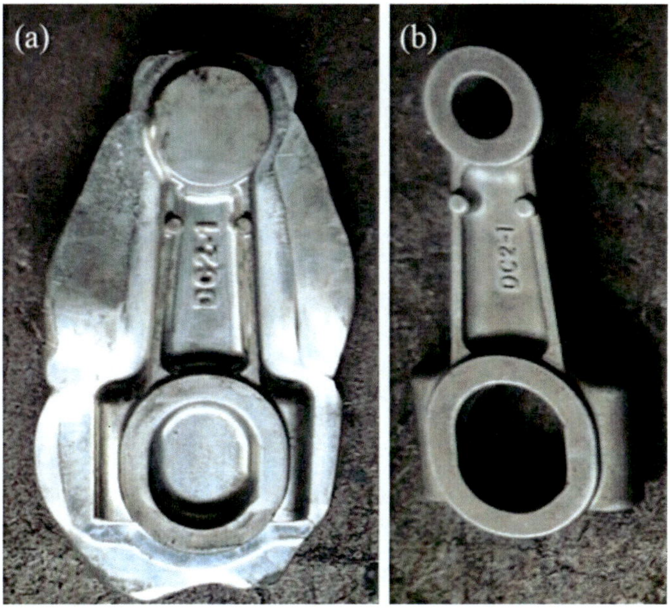

Fig. 4.35 2014 aluminum alloy connecting rod: **a** final forging with flash and punching wad, **b** forging after trimming, punching and sizing

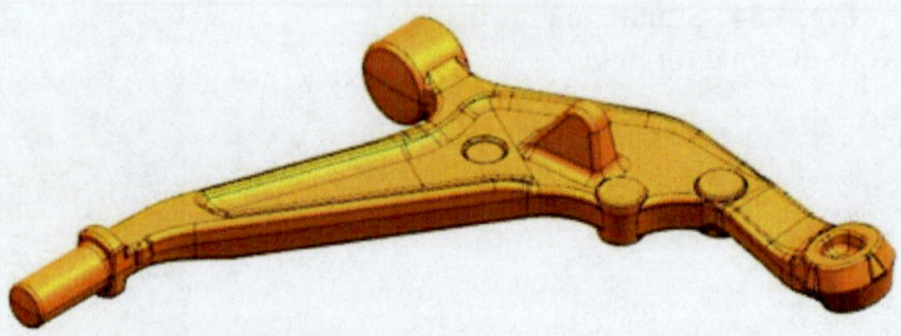

Fig. 4.36 A typical branch-shaped aluminum alloy control arm

was cutting → billet-making → bending → pre-forging → final forging. The cross-sectional area of the control arm changes significantly along the axis direction. Therefore, the amount of metal transfer is very large, but it is difficult to achieve good results by die forging for billet preparation. The roll forging is very suitable for preparing the billets whose cross-sectional area changes significantly.

During the design of roll forging process, the billet was divided into several characteristic sections, such as head and rod, according to the calculated billet drawing. The curve of the calculated billet was replaced by a straight line under the principle of volume constant. The length of the end section of the billet was slightly shorter than that of the forging, so that the billet was easier to put into the forging die. The length of each middle section was the same as the length of the corresponding section on the forging.

A transition region was required between each section. The slope of the transition region was 45°–60°. The length of the transition region was calculated according to the Eq. (4.7). The size of the original bar for roll forging was selected according to the largest section. The roll forged part designed according to the above rules is shown in Fig. 4.37.

Taking into account the structural characteristics of the control arm, a bar was selected as the billet of roll forging, so the most commonly used elliptical-circular groove system was selected. The minimum cross-sectional diameter of the roll-forged part was ϕ 34 mm, and the cross-sectional diameter of the non-deformed section in the middle was ϕ 65 mm. Therefore, the total elongation was 3.65. The average elongation of the ellipse-round groove system generally does not exceed 1.5, so the calculated roll forging pass was 4.

There was a long non-deformed section in the middle of the roll forged part. Two process schemes could be employed. One is U-turn roll forging, that is, to roll the right side of the part shown in Fig. 4.37 first, and then take the right side as the

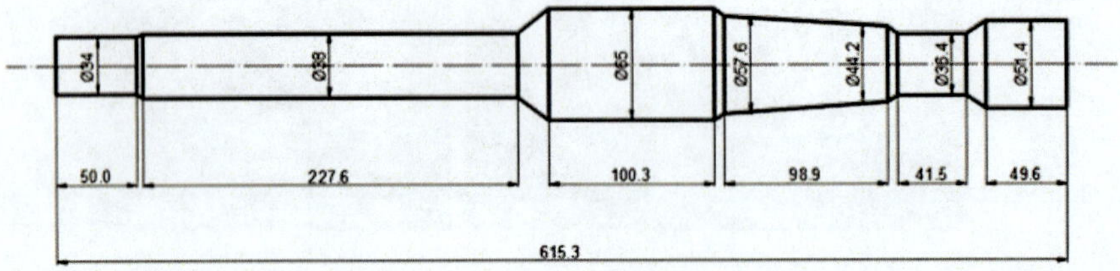

Fig. 4.37 The roll forged part

clamping end, and roll the left side. The other is to set a clamping head at one end of the part, so that it can be rolled sequentially, but material is wasted.

With the widespread application of automation devices, it is relatively easy to realize a U-turn in actual production, so the U-turn roll forging was used. The program of roll forging is shown in Fig. 4.38.

During the design of bending process, the roll forged billet does not require significant elongation and deformation, so the free bending process was employed. The bending corners needed to be sufficiently large to prevent the billet from folding defects during the die forging process. When designing the bending dies, the lower die was set with two supporting points to keep the billet horizontal state when it was placed in the die.

The main part of the forging was in a horizontal plane, and the curved arm on the right was an upwardly inclined surface. Therefore, it was necessary to use several planes to divide the die, and the different planes were smoothly transitioned through rounded corners. The parting surface is shown in Fig. 4.39.

The model of the final forging die is shown in Fig. 4.40.

During die forging, the metal flowed severe along the horizontal direction and smoothly along the height direction. Therefore, when designing the pre-forging die cavity, the height was greater than that of the final forging die cavity, and the width was slightly smaller than that of the final forging die cavity. The round section of the final forging was designed as an elliptical section during pre-forging. The metal volume was allocated reasonably to avoid defects such as folding and back-flow. The structure of the pre-forged part was conducive to its positioning in the final forging die. The pre-forging die is shown in Fig. 4.41.

2. Numerical simulation of forging process

The simulation result of the pre-forging process is shown in Fig. 4.42. It can be seen that the transition between the middle section and the long wedge section is smooth, which avoids the occurrence of folding defects. The metal flow situation at the end of the pre-forging shows that the metal flows rapidly at the branch on the left and the flash groove on the right, and the amount of metal at the branch is sufficient to ensure accurate forming during the final forging.

In order to improve the filling capacity in the branch during the final forging, a damping groove was set in the final forging die. The simulation result of the final forging is shown in Fig. 4.43. It can be seen that due to the effect of the damping groove, the flash near the branch has been reduced, the filling has been improved.

The finite element simulation results indicated that the designed process plan was feasible. The forged control arm is shown in Fig. 4.44.

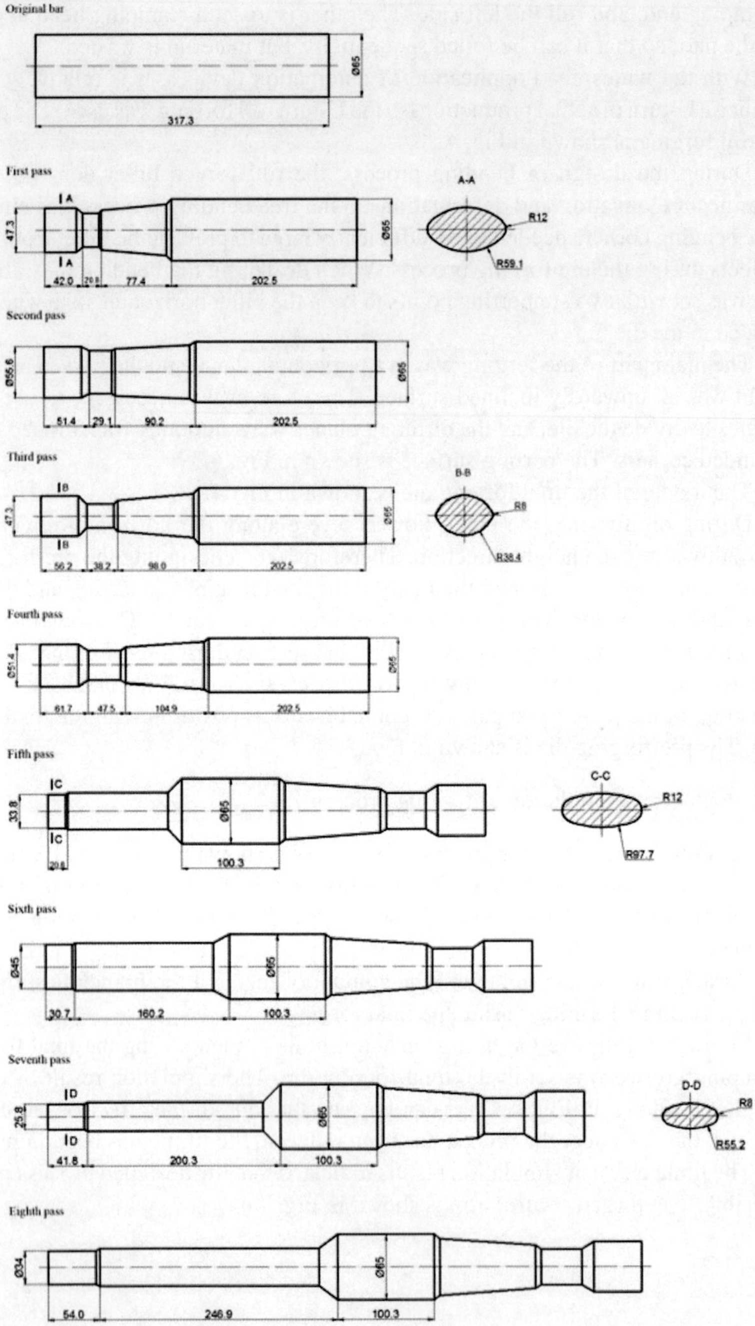

Fig. 4.38 Roll forging program

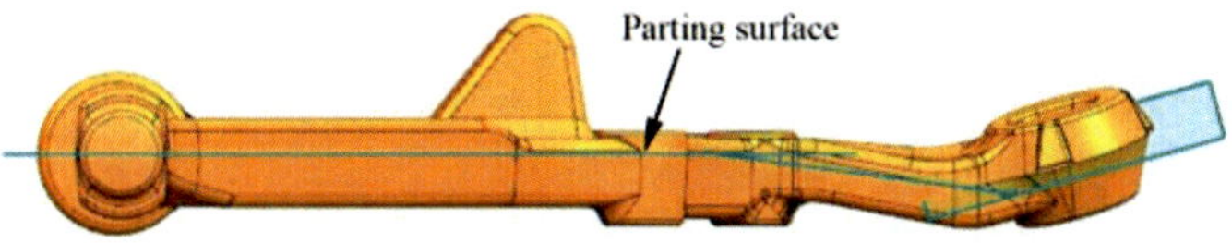

Fig. 4.39 Parting surface of the final forging

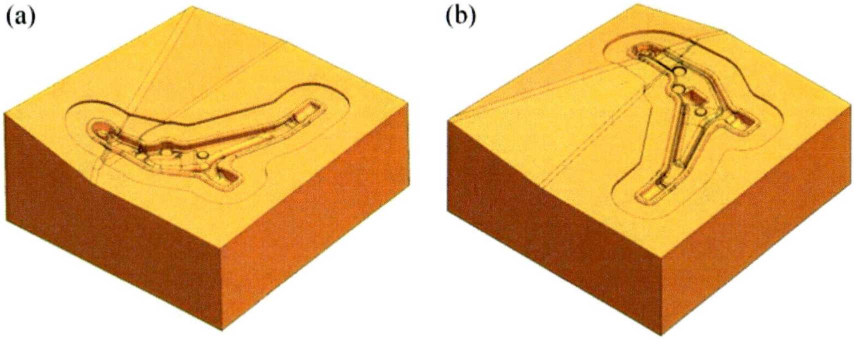

Fig. 4.40 The final forging die: **a** upper die, **b** lower die

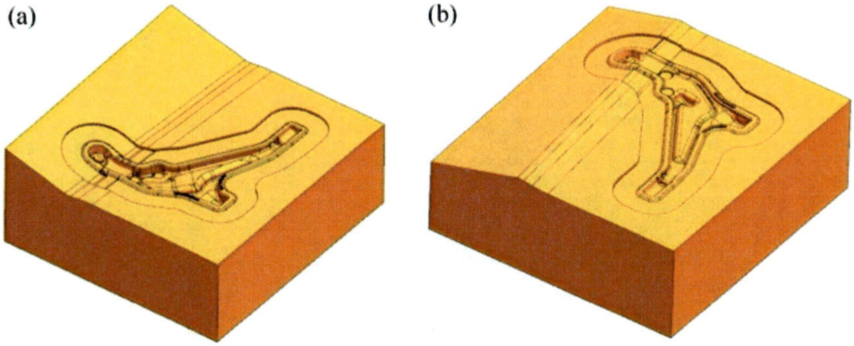

Fig. 4.41 The pre-forging die: **a** upper die, **b** lower die

4.8.3 Die Forging of 6082 Aluminum Alloy Wingspan Control Arm

1. Die forging process

The three-dimensional shape of the wingspan control arm is shown in Fig. 4.45. Its shape is similar to the above-mentioned control arm. It is a curved part with ribs on the periphery, and porous thin webs between the ribs. Since the length is much larger than the width and thickness, it was treated as a long shaft forging when designing

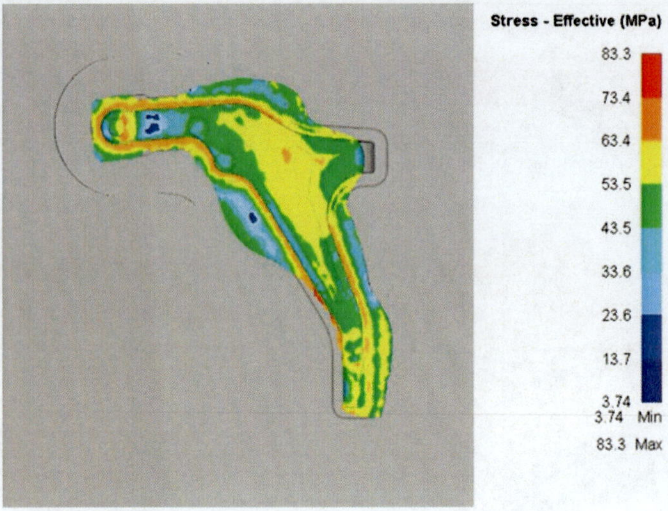

Fig. 4.42 Filling situation of the pre-forged part

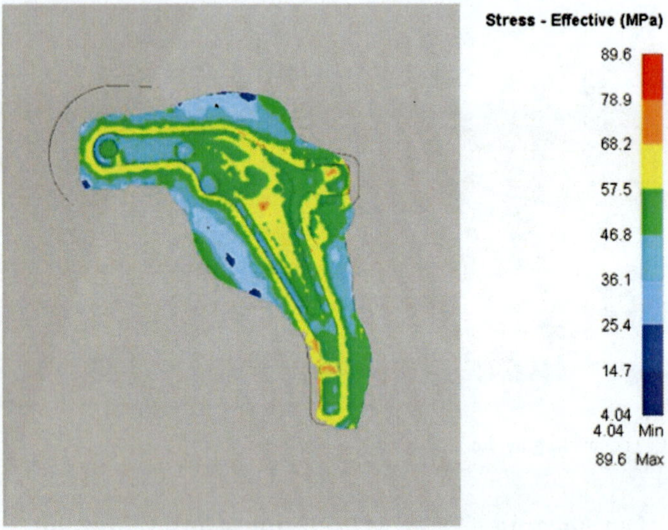

Fig. 4.43 Filling situation of the final forged part

the die forging process. The designed process route was cutting → heating → roll forging → bending → flattening → second heating → pre-forging → final forging → trimming, punching, and sizing [8].

According to the design results of the roll forging process, the modeling of roll forging module was established, and then the Boolean operation was used to generate

Fig. 4.44 The forged control arm

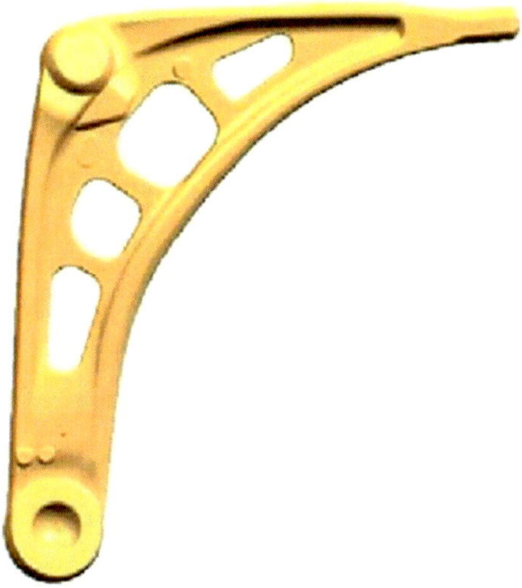

Fig. 4.45 Three-dimensional modeling of wingspan control arm

the cavity through the modeling methods such as sweeping, stitching, supplementation and shearing. The transition line was turned into a fillet. Finally, the modeling of four-pass roll forging die was obtained, as shown in Fig. 4.46. The manufactured die for roll forging is shown in Fig. 4.47.

According to the three-dimensional model of the roll forgings, the bending die was designed by bending along the bending line and extending the curved surface of part of the die cavity. The manufactured bending die is shown in Fig. 4.48.

The three-dimensional model of the wingspan control arm forging die and the actual forging die are shown in Fig. 4.49 and Fig. 4.50 respectively. The upper part of the die is the pre-forging cavity and the lower part is the final forging cavity.

Figures 4.51 and 4.52 show the three-dimensional modeling of trimming-punching-sizing compound die and the actual die, respectively, which is a compound die with three functions of trimming, punching and sizing. In one stroke of the slider,

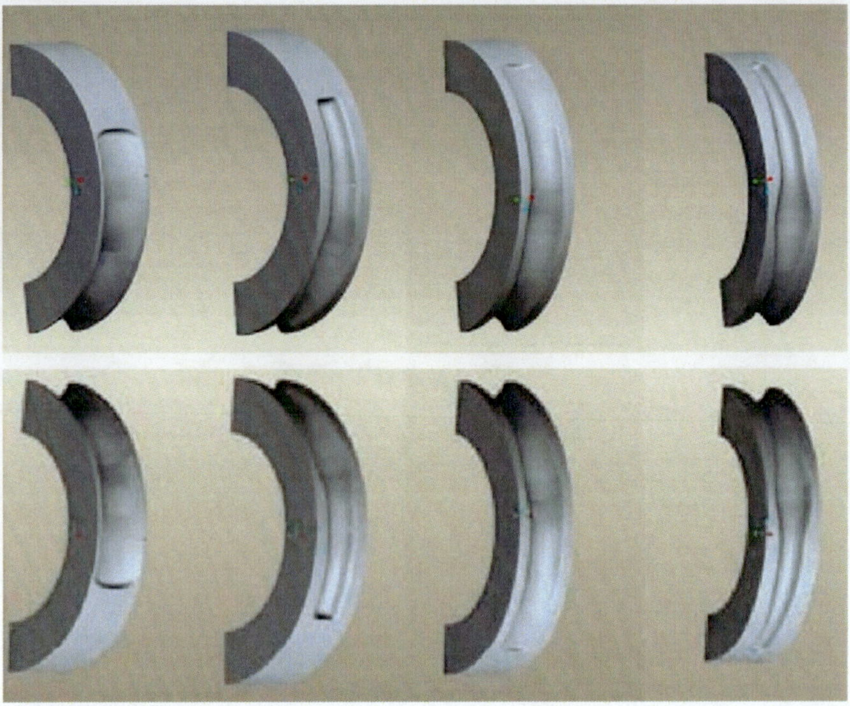

Fig. 4.46 Three-dimensional modeling of the roll forging die

Fig. 4.47 The manufactured roll forging die

Fig. 4.48 The bending die

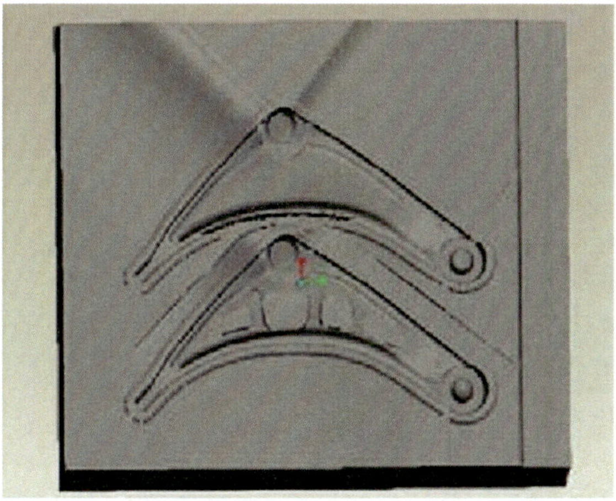

Fig. 4.49 Three-dimensional model of the forging die

the trimming, punching, and sizing were performed. Compared with the single-step die, its production efficiency is high.

2. Process test

On the basis of theoretical analysis and finite element simulation results, a process test was carried out. The actual samples of the roll forged part, bent part, flattened part, pre-forged part and final forging are shown in Fig. 4.53. It can be seen that the forgings are fully filled, and there are no defects such as folding, and underfilling.

Fig. 4.50 The actual forging die

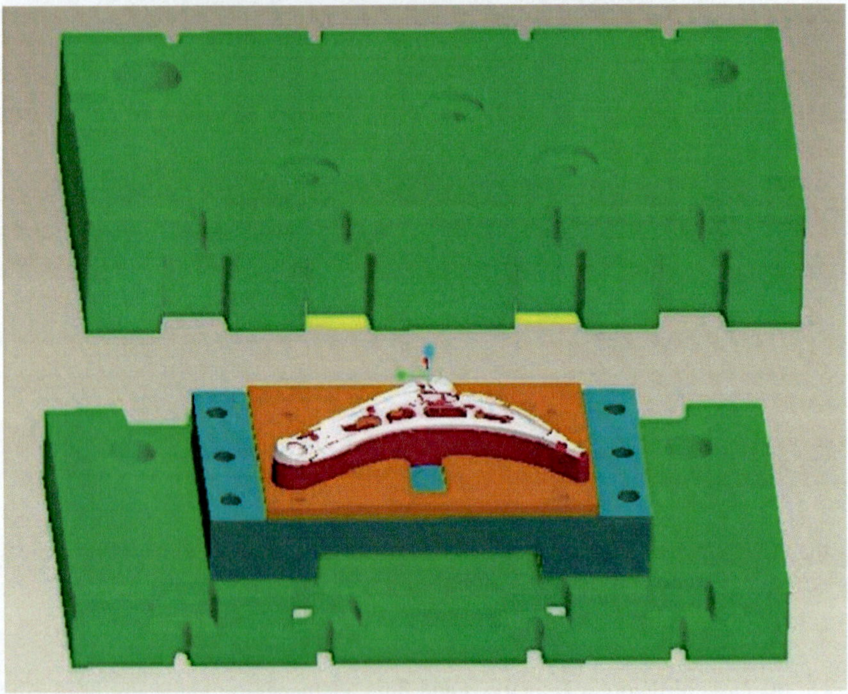

Fig. 4.51 Three-dimensional model of the trimming-punching-sizing compound die

Fig. 4.52 The trimming-punching-sizing compound die: **a** upper die, **b** lower die

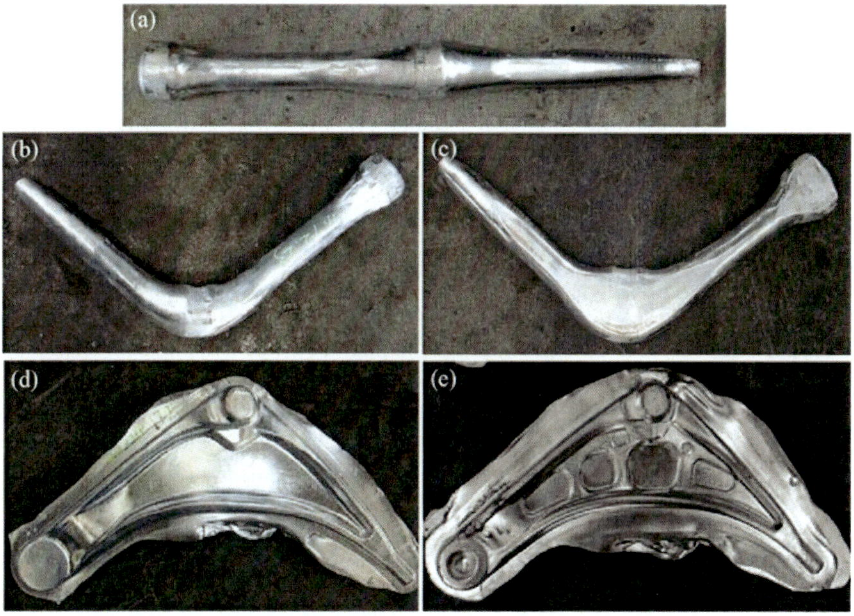

Fig. 4.53 Forging process of wingspan control arm: **a** roll forged part, **b** bent part, **c** flattened part, **d** pre-forged part, **e** final forging

4.8.4 Multi-directional Precision Forging of 7075 Aluminum Alloy Casing

1. Precision forging process

The casing is a key part of the firearm, and made of 7075 aluminum alloy. It can be seen from Fig. 4.54 that the casing is a long rod with U-shaped grooves, which has obvious structural characteristics of long shaft forgings. Therefore, the precision

Fig. 4.54 Three-dimensional model of the casing

forging process was designed according to plane deformation [9–11]. This book introduces its multi-directional precision forging process and optimization in detail.

The forging process route was cutting → heating → billet-making → pre-forging → final forging → sandblasting → deburring.

In order to ensure accurate cutting, a horizontal high-velocity automatic band sawing machine was used. This cutting method has flat fractures and no defects such as collapsed corners. A catenary pass-through heating furnace was used for heating. This kind of heating furnace not only has uniform heating, but also has a driving device which is easy to realize automation and continuous production. Before forging, the billet was kept at a temperature of 430 °C for 1 h. The catenary pass-through heating furnace has a long track which can support multiple batches of heating at the same time. Through real-time or intermittent transmission, continuous production can be achieved.

The principle of the hot extrusion process to make billets is shown in Fig. 4.55. The heated bar was placed in the cavity of the die, and then the left punch and right punch were opposed to each other to force the billet to be formed into the required shape. The separable dies were used for taking out the billet.

The hot extrusion process was carried out on a multi-directional die forging press. During extrusion, the material was deformed in a state of three-dimension compressive stress state which is conducive to forming. The billet obtained by hot extrusion had accurate dimensions and good surface quality. Different sizes and shapes of

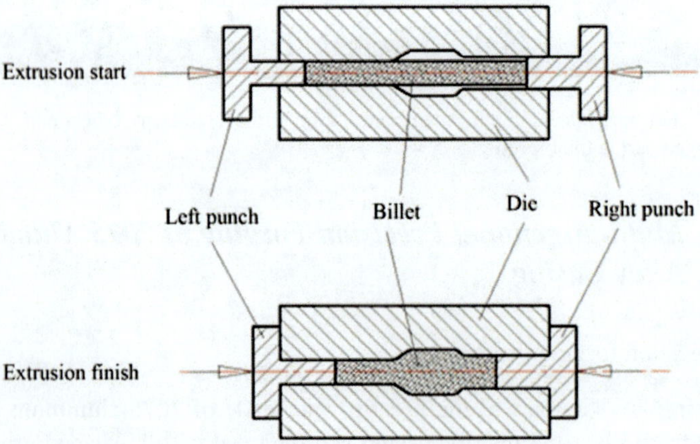

Fig. 4.55 Schematic of hot extrusion for making billet of casing

billets can be produced by changing the die. Compared with roll forging and cross wedge rolling, although the production efficiency of hot extrusion was slightly lower, multiple cavities could be designed on a pair of dies to obtain multiple parts in one die and improve production efficiency.

Pre-forging and final forging were completed in two steps. The pre-forging die cavity and the final forging die cavity were arranged side by side on one die. The final forging was directly performed by manual overturning after the pre-forging was completed. This can not only shorten the process flow and improve production efficiency, but also greatly reduce production costs by reduction of equipment and dies.

After die forging, the surface of the forging needs to be cleaned to remove the lubricant and burrs on the surface, which mainly includes two steps: sandblasting and deburring. Dry sandblasting machine and liquid sandblasting machine can be used for sandblasting. Because the casing is a mass-produced part, a liquid sandblasting machine was selected to reduce dust pollution and improve the operating environment. Since the burrs of the casing forgings were small and thin, no special edge trimming equipment was required, and only manual deburring was required.

2. Two-step multi-directional precision forging

There were two parting schemes of dies for the multi-directional precision forging of the casing: vertical parting and horizontal parting, as shown in Fig. 4.56 and Fig. 4.57, respectively.

At the beginning of forging on the vertically parting die, the left die and the right die moved toward each other into a half-closed state, and the heated billet was placed in the half-closed die. Since the bottoms of the left and right dies are "L"-shaped, the billets were contained without falling out of the die. Then the two dies were closed to form an integral die while the billet was flattened. The slider of press drove the punch to go down, and the billet was squeezed into a casing forging. After forging, the punch returned, the left die and right die were opened, and the forging was taken out.

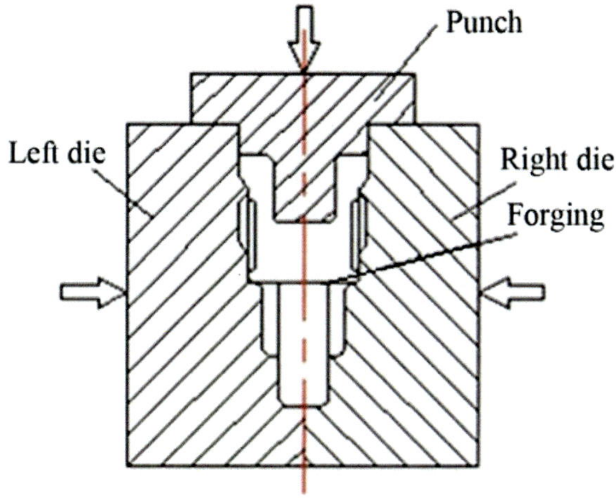

Fig. 4.56 Schematic of vertical parting die

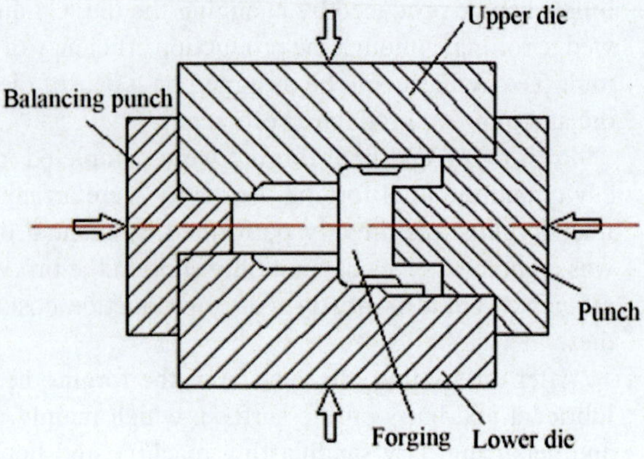

Fig. 4.57 Schematic of horizontal parting die

At the beginning of forging on the horizontally parting die, the heated billet was placed on the lower die, the slider of press drove the upper die to go down. The upper die and lower die were closed into an integral die and the billet was flattened. Then, the balancing punch moved to the right to the extreme position. The punch moved to the left and forced the billet to be squeezed into the casing forging. After forging, the punch and balancing punch returned, and the slider drove the upper die to return to the upper limit position, the forging was taken out.

Comparing the two schemes, the vertical parting is more suitable for one-step die forging; the horizontal parting allows multiple steps of production to be completed on one press, and its efficiency is high. In addition, because the lateral projection area of the casing was larger than the horizontal projection area, that is, a greater forming force was required in the lateral direction of the casing, the horizontal parting was more suitable for taking advantage of the large tonnage of the main slider of the press. Therefore, the horizontal parting scheme was used for the multi-directional precision forging of the casing.

The die of the casing was designed as a vertically parting structure, as shown in Fig. 4.58. The die was divided into the following parts: upper die, lower die, pre-forging punch, final forging punch, and guide key. The punch and the dies were used to form a closed die cavity to form the forging. The guide key had two functions. One was to ensure that the upper and lower dies are in good position in the clamping stage, and the other was to combine the upper and lower dies in the extrusion stage to jointly bear the forging force.

3. Numerical simulation of forging process

The geometric model of numerical simulation was established based on the actual size of the casing. In order to reduce the amount of calculation, the deformation of the die was not calculated, both the punch and the die were set as rigid bodies. The thermal–mechanical coupled rigid viscoplastic finite element method was used for numerical simulation of the forging process.

Three-dimensional tetrahedral grids were used to mesh the billet and the die. The grid number of billet was 100,000, the grid number of die was 150,000, and the grid

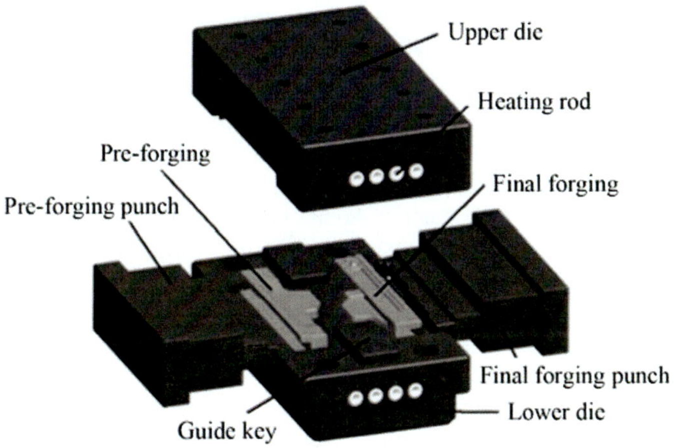

Fig. 4.58 Diagram of multi-directional precision forging die

number of punch was 8000. The material of billet was 7075 aluminum alloy, and the material of die was H13 steel. The initial temperature of the billet was 430 °C, the temperature of the die was 200 °C. The friction factor was 0.3, and the punch velocity was set to 30 mm/s.

The strain distribution during the forging process is shown in Fig. 4.59. The forming process of the casing was divided into pre-forging and final forging. In the

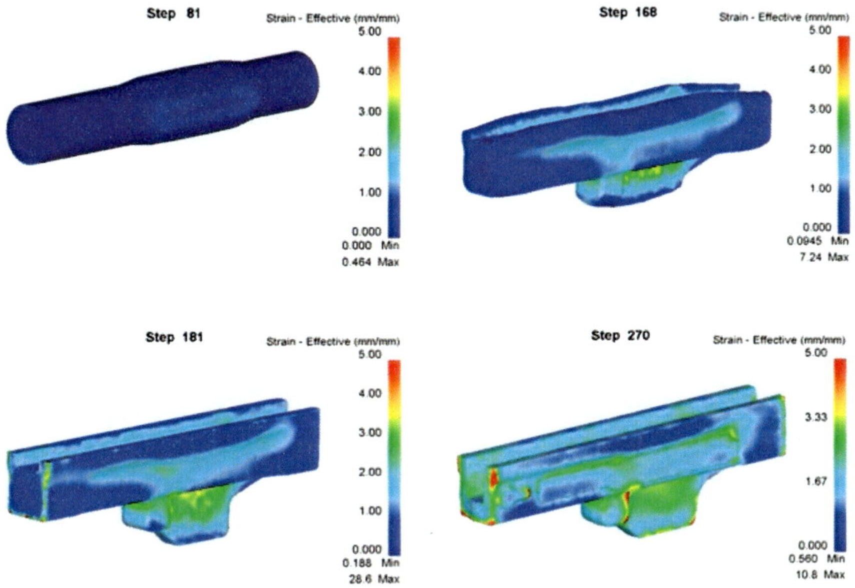

Fig. 4.59 Cloud diagram of strain distribution

clamping stage of pre-forging, the part of the billet with the larger middle diameter was flattened. With the action of the pre-forging punch, the billet gradually deformed into a shape similar to the casing. The detailed features of the forging were formed in the final forging process. In order to ensure the "fullness" of the special-shaped bosses and slender ribs, a large forging force and clamping force were required to make the forgings in a severe three-dimension compressive stress state and force the metal to flow to the corners, gaps and other parts that are difficult to be filled.

It can be seen that the deformation of the material is mainly concentrated on the lower rib and U-shaped groove during the pre-forging process. The lower rib was formed by forward extrusion, and the U-shaped groove was formed by backward extrusion. The deformation of the material during the final forging was mainly concentrated in the detailed parts such as special-shaped bosses, slender ribs, multiple steps and the edge of the forging. The deformation of the forging during the final forging process was small, but it played a decisive role in the quality of the forging.

The temperature distribution of the forging and the die during the forming process are shown in Fig. 4.60 and Fig. 4.61, respectively. It can be seen that due to the heat transfer between the forging and the die, the temperature of the forging gradually decreases as the deformation progresses. But the temperature in the regions with greater deformation decreases more slowly, and even the temperature of some regions increases. This is due to the transformation of plastic deformation work and friction work into heat energy which offsets part of the heat loss. It can be seen from Fig. 4.61 that the temperature of the die rises rapidly. A higher temperature was conducive to

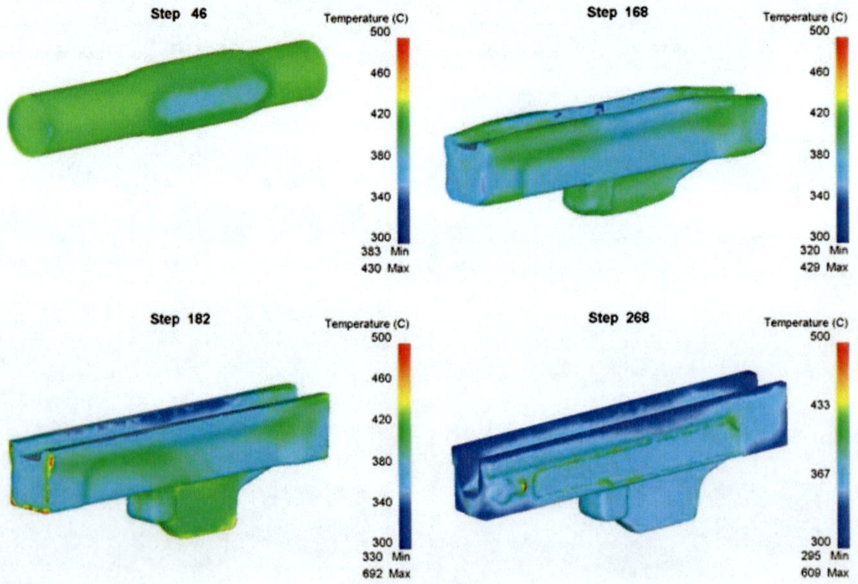

Fig. 4.60 Temperature distribution of the forging

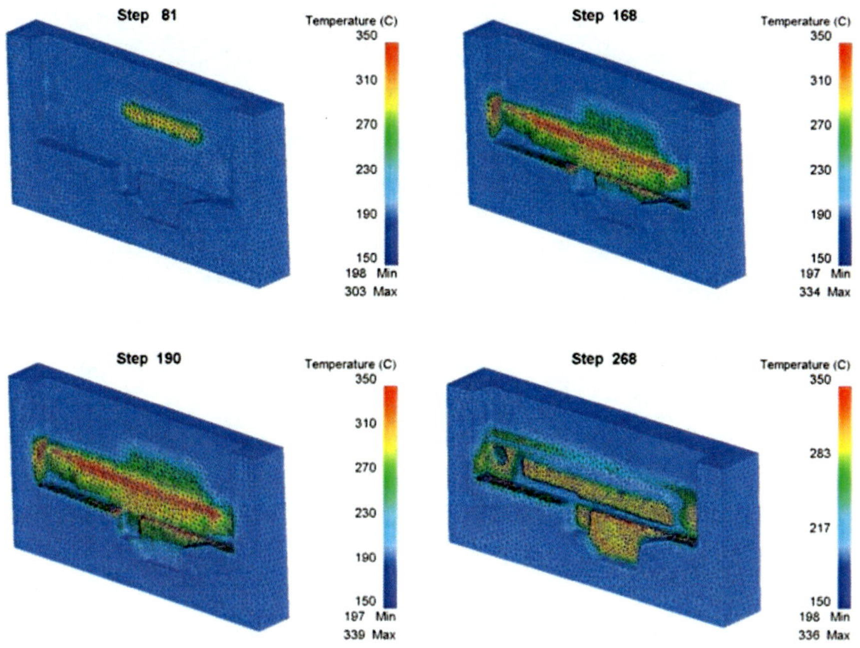

Fig. 4.61 Temperature distribution of the die

the flow of materials, but at the same time, the temperature of the die should be controlled not to be too high to prevent softening and rapid wear of the die.

4. Process test

On the basis of theoretical analysis and finite element simulation, the multi-directional precision forging was carried out. The raw material was extruded 7075 aluminum alloy bar. Before the precision forging, the billet was kept at a temperature of 430 °C for 1 in a catenary pass-through heating furnace. The assembled die shown in Fig. 4.62 was heated to 250 °C. It was installed on a YK34J-1600/C1250 multi-directional hydraulic press. The lower die was fixed on the worktable, the upper die was connected to the slider of the press. The pre-forging punch and the final forging punch were connected to the left and right extrusion sliders, respectively. The lower die was equipped with ejector rod.

The forgings of casing produced by multi-direction forging process are shown in Fig. 4.63. The outline of the forgings was clear, and both sides and bottom met the dimensional accuracy requirements of the parts.

Compared with traditional hammer forging, the number of heating was reduced from 3 to 1. The material utilization was increased from 48 to 80–85%. Compared with the isothermal forging, the material utilization was increased from 74 to 80–85%, which greatly saved heating energy consumption and increased die life.

Fig. 4.62 The multi-directional forging die for casing

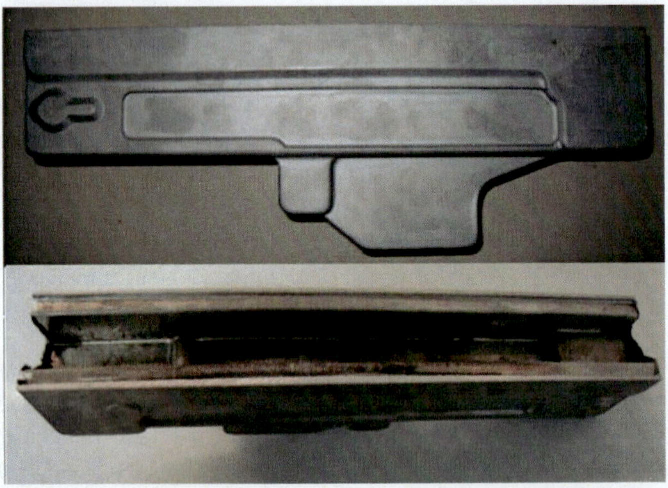

Fig. 4.63 The forgings of casing

4.8.5 Small Flash Precision Forging of 6082 Aluminum Alloy Curved Control Arm

1. Die forging process

The three-dimensional modeling of a typical curved control arm is shown in Fig. 4.64. The designed small flash precision forging process was as follows: cutting → heating → bending → small flash precision forging → trimming → punching → sizing [12].

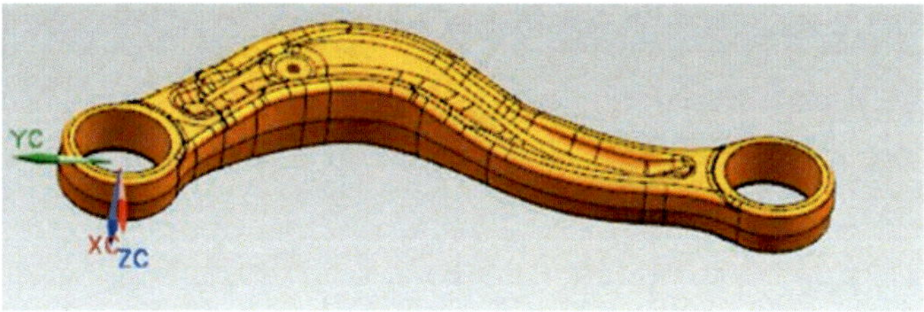

Fig. 4.64 Three-dimensional modeling of the curved control arm

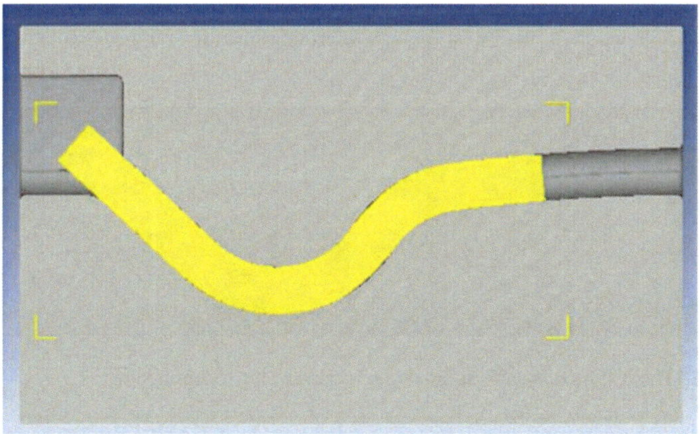

Fig. 4.65 Simulation results of the bending

2. Numerical simulation of forging process

The alignment between the bent part and the target part are shown in Fig. 4.65. There was a certain difference between the actual bent part and the target part, but it is very small and does not affect the final forging. The temperature range of the bent part was about 450–500 °C. Except that the part in contact with the die was reduced by about 50 °C, the temperature of the part was still within the reasonable forging temperature range.

The filling condition of the typical cross section of the forging at the end of the small flash precision forging is shown in Fig. 4.66. It can be seen that the forging is good, which is because the small flash increased the resistance of the metal to flow into the flash groove laterally, and improved the filling of the cavity in the longitudinal direction. It can also be seen from Fig. 4.66a that the flash is small, which is beneficial to improve the material utilization. The temperature of the forging was 450–475 °C which is a few lower than the initial temperature due to the fact that the velocity of the press reaches 700 mm/s, and the heat generated by the deformation made up for the heat loss. The simulation results verified that the designed small flash precision forging process is reasonable.

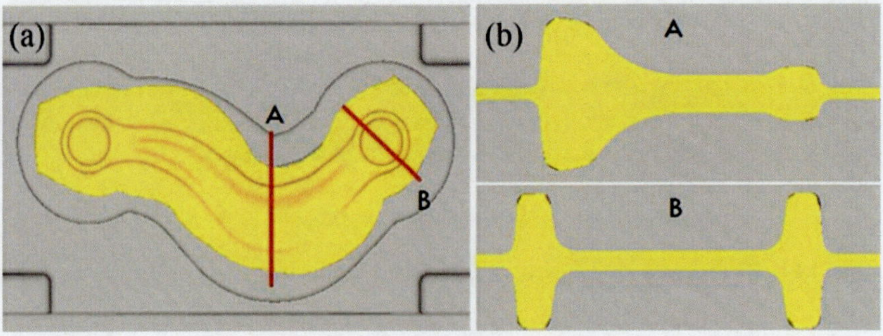

Fig. 4.66 Simulation result of final forging: **a** forming situation, **b** typical sections of forging

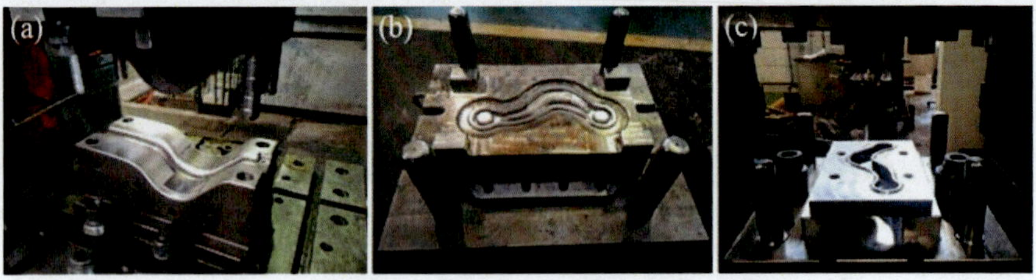

Fig. 4.67 The testing dies: **a** bending die, **b** forging die, **c** trimming die

Fig. 4.68 Forgings of control arm

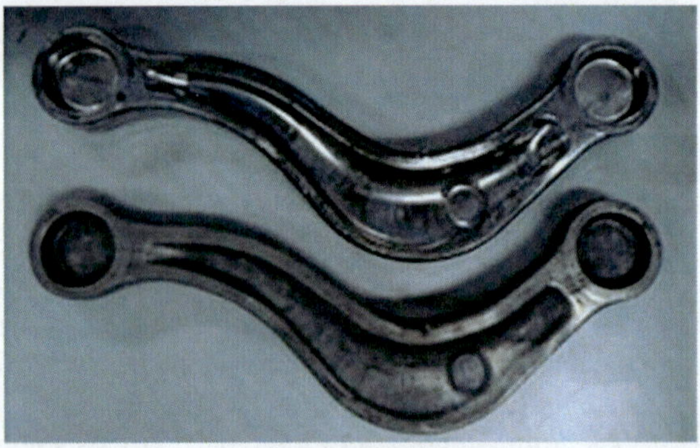

3. Process test

The manufactured bending die, forging die and trimming die (Fig. 4.67) were installed on a 4000 kN hydraulic press, a 6300 kN friction press and a 2000 kN hydraulic press, respectively. The forging of control arm is shown in Fig. 4.68. It can be seen that the forging is fully formed, which indicates that the designed process is feasible.

References

1. Wu SX, Pan QJ (2014) Handbook of wrought aluminum alloy and its die forging technology. China Machine Press, Beijing
2. Xia JC, Zhang QX (2010) Material forming technology. China Machine Press, Beijing
3. Zhang YJ, Chen TF, Yang J, Xia JC, Deng L, Jin JS (2018) Development and application on small flash precision forging technology for atuo steering knuckle. Forigng Stamp Technol 43(8):1–7
4. Zhang ZW (1988) Forging technology. Machinery Industry Press, Beijing
5. Wei W, Jiang P, Cao F (2012) Blank-making and roll forging process and design of die for aluminum controlling arm. Die Mould Ind 38(5):61–65
6. Wei W, Jiang P, Cao F, Yang Y (2012) Synchronization of the movement between blank and manipulator in automatic roll forging. J Plast Eng 19(1):11–15
7. Xiao RQ (2015) Research on die forging technology of complex aluminum alloy steering arm. Huazhong University of Science and Technology, Wuhan
8. Wei W (2013) Research on hot deformation behavior and microstructure property of 6082 aluminum alloy forging with rib. China Academy of Machinery science and Technology, Beijing
9. Li QJ (2011) Optimization of multi-directional precision forging for high strength aluminum alloy casing. Huazhong University of Science and Technology, Wuhan
10. Deng L, Xia JC, Wang XY, Hu GA (2009) Multi-directional precision forging for casing. China Mech Eng 20(7):869–871
11. Li QJ, Xia JC, Deng L, Chen XZ, Hu Z, Fan QS (2010) Optimization of multi-directional precision forging for aluminum alloy casing. Forigng Stamp Technol 35(5):24–28
12. Ren ZH (2018) Research on shape and performance control of 6082 aluminum alloy long axis forgings with rib-web section. Huazhong University of Science and Technology, Wuhan

Chapter 5
Precision Forging Technology for Complex Revolving Parts

5.1 Overview

The structural characteristics of the revolving aluminum alloy forgings are based on a simple axisymmetric revolving body, and some have bosses or pits at the top or bottom or both ends; some have branch or pits on the side; some are multilayer thin-walled cylindrical structures with different wall thicknesses and heights.

Generally, the complex revolving aluminum alloy forgings can be formed with integral closed precision forging or separable closed precision forging. The process principle of closed precision forging is that the aluminum alloy billet is formed into the required forging by the force of the punch in the closed die cavity composed of punch and dies. Therefore, closed precision forging can make the geometry, dimensional accuracy and surface quality of forgings as close as possible to finished parts.

Compared with open die forging, closed precision forging has the characteristics of higher material utilization, higher production efficiency, lower consumption of heating energy, better quality of forgings, and reduced machining cost. The closed die forging, especially the separable closed precision forging, does not generate flash, and the forging slope is 1°–3° or even without slope, and pits perpendicular to the forging direction can be forged. These advantages can increase the material utilization by more than 20% on average. The use of separable die precision forging can often reduce the billet-making step and the number of die forging steps by 1–2, and avoid the trimming step. Therefore, the production efficiency can be increased by 25–50% on average. It is easier to realize the automation of precision forging production. In addition, the saving of heating energy consumption is accompanied by the improvement of material utilization.

On the one hand, the closed precision forging can make the shape of the forging and the finished part very close or completely consistent, so that the metal flow line is continuously distributed along the outline of the finished part. On the other hand, the deformed metal is in a state of three-directional compressive stress, which can inhibit

L. Deng et al., *Precision Forging Technology and Equipment for Aluminum Alloy*, Springer Series in Advanced Manufacturing,
https://doi.org/10.1007/978-981-19-1828-5_5

the occurrence of cracks in the part. Thus, the mechanical properties of the product are relatively higher than that of general open die forgings. In addition, since the forgings have no flash, there will be not the exposure of flow line which is generated by trimming. This is beneficial for parts with high corrosion resistance requirements, especially for aerospace and marine aluminum alloy parts.

Closed precision forgings not only have a small cutting allowance, but also have no residual flash left when trimming the flash, which can effectively reduce subsequent machining and reduce tool consumption, thereby helping to save machining costs.

At present, the closed precision forging with separable die is mainly developed along two technical routes. One is the combined die structure composed of a general die base and replaceable punch and die, which is used in general forging equipment such as the mechanical hot forging press or crank press, hydraulic press and screw press to realize the precision forging of some medium and small forgings. The other is to use special equipment, such as mechanical, hydraulic or mechanical-hydraulic combined double action and multi-directional precision forging presses to realize the precision forging of various complex forgings.

High-strength aluminum alloys is characteristic of high strength and poor plasticity. During closed precision forging, not only the forging force increase rapidly, but also some narrow and deep parts in the die cavity are extremely difficult to fill. To solve this problem, a new flow control forming process was studied. The principle is that a gradually decreasing compressive stress gradient field is formed from the entrance of the cavity to the most difficult filled part of the cavity by setting different control methods to ensure that the most difficult filled part is also filled when the other parts of the cavity are filled [1].

5.2 Closed Precision Forging of 7075 Aluminum Alloy Gland and Housing

5.2.1 Design of Forging Process

Figure 5.1 shows the precision forgings of the gland and the housing of the airbag gas generator. The material is 7075 aluminum alloy. It can be seen that the forgings are multi-layer cylindrical structures. The difference is that there is a round boss with a diameter of ϕ19 mm in the middle of the gland and a flange at the bottom.

According to the structural characteristics of the two forgings, either forward extrusion or backward extrusion can be used. The extrusion principle of the gland is shown in Fig. 5.2. The forming schemes of the housing is similar. It is necessary to make a judgment through analysis to determine which kind of extrusion process to use. If the forward extrusion process shown in Fig. 5.2b was used, the punch had a simple structure, the die employed a radial layered combination structure, and meanwhile, a thin-walled cylindrical ejector should be provided. Since the ejection stroke was designed according to the maximum height of the forging, the multi-layer

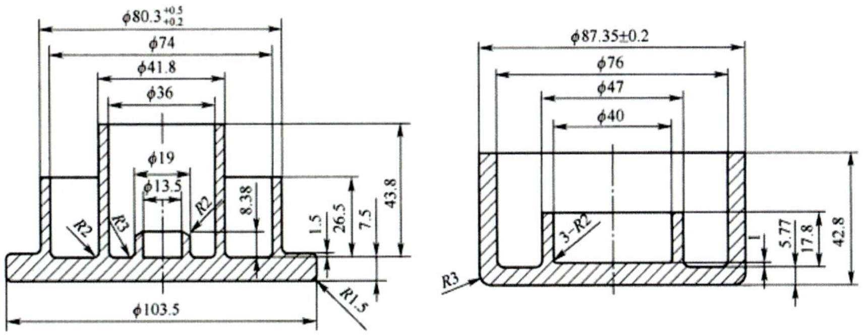

Fig. 5.1 Diagram of gland and housing

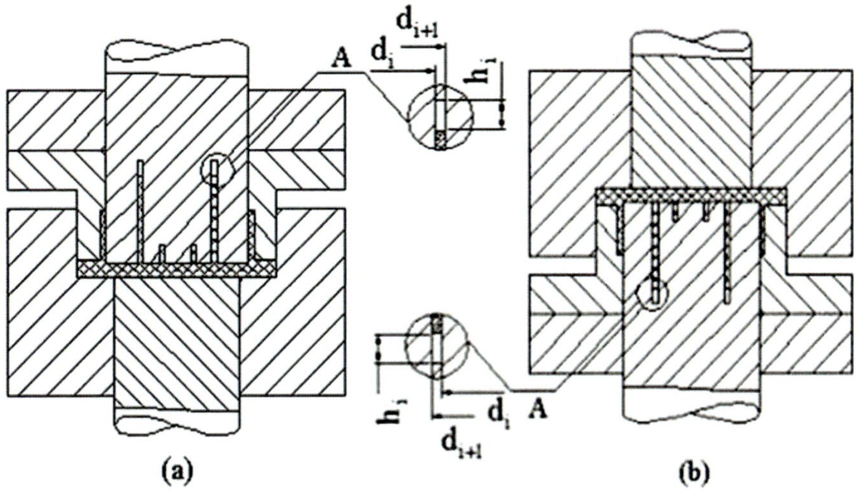

Fig. 5.2 Schematic diagram of backward extrusion (**a**) and forward extrusion (**b**)

cylinder cavity of the combined die was narrow and deep, which not only makes the structure of the die complex, but also makes it difficult to eject the forging. If the backward extrusion process shown in Fig. 5.2a was used, the structure of the die was simple, and the punch employed a radial layered combination structure. At this time, the outer layer of the combined punch can be used as a stripper, and the multi-layer cylinder cavity was set in the combined punch, and the depth of the cavity only needed to be designed according to the height of the forging. The later scheme greatly reduces the complexity of the die structure, and the forging can be easily removed from the combined punch. It should be pointed out that the backward extrusion force is greater than the forward extrusion force. From the above analysis, the backward extrusion was selected.

Whether the forging is formed at one time can be judged by calculating the deformation degree of extrusion. For the gland, the deformation degree was 83.8% which is less than the allowable deformation degree of 7075 aluminum alloy (90–95%), indicating that the gland can be formed by one-step backward extrusion.

After selecting the backward extrusion forming process, cold backward extrusion and hot backward extrusion were compared. Through finite element simulation and experimental research, it was found that during cold backward extrusion, qualified forgings could be obtained, but the extrusion force was significantly increased, which was more than 4 times the hot extrusion force. Meanwhile, the thickness of the bottom of the forging should be increased by 2–3 mm compared with that of the hot forging to realize forming. Due to the poor plasticity of the material, obvious microcracks appeared in the lower section of each layer of the cylinder of the forging, but still within the machining allowance. On the whole, because of the low forging force of hot backward extrusion, high material utilization, and lower possibility to generate cracks, the hot backward extrusion process was selected as the forming process of the gland and the housing.

5.2.2 Design of Flow Control Chamber

Through the design of control chamber, the forging has a free surface at the last filled part, and a stress gradient can be formed from the entrance of the backward extrusion cavity to the most difficult filled part, causing favorable conditions for metal to flow there [2].

If the control chamber is set at the end of the cavity where the metal is last filled in the cylindrical cavity of the die, a judgment must be made first, that is, for the cavities with different thicknesses and heights, which cavity is the last to be filled?

According to the law of minimum flow resistance in plastic forming and experimental observations, the final filled part is mainly judged by the flow resistance of the die cavity to the deformed metal. For this reason, a criteria was proposed to judge the final formed part by the ratio k_{i} of the height to width of the cylinder wall and the ratio m_{i} of the cross section of the cylinder wall $S_{\mathrm{i}}^{\mathrm{p}}$ to the projected area of the corresponding deformation area $S_{\mathrm{i}}^{\mathrm{f}}$, namely

$$k_{\mathrm{i}} = h_{\mathrm{i}} / t_{\mathrm{i}} \tag{5.1}$$

$$m_{\mathrm{i}} = S_{\mathrm{i}}^{\mathrm{p}} / S_{\mathrm{i}}^{\mathrm{f}} \tag{5.2}$$

where h_{i} is the height of the cylinder wall; t_{i} is the width of the cylinder wall; $S_{\mathrm{i}}^{\mathrm{p}}$ is the cross-sectional area of the cylinder wall; $S_{\mathrm{i}}^{\mathrm{f}}$ is the projection area of the corresponding deformation area of the cylinder wall.

The larger the value of k_{i} and m_{i}, the more difficult it is to form the cylindrical wall. When the dimensional accuracy and surface roughness of all cylindrical cavities

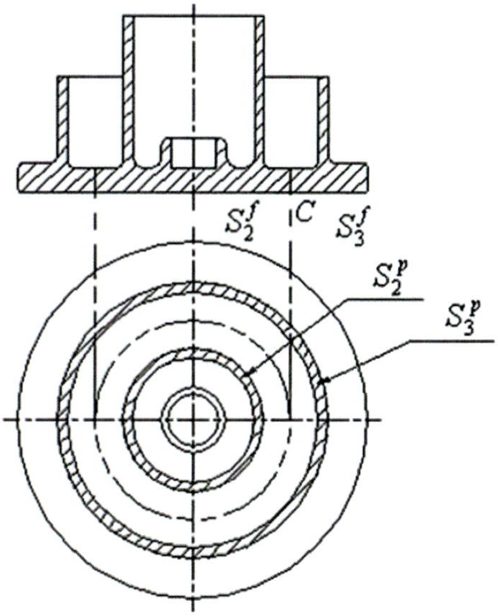

Fig. 5.3 Schematic of the splitting surface of the gland

are the same, the narrower and deeper the cavity, the greater the resistance to metal flow. In addition, the smaller amount of metal in the vicinity of the die cavity makes it difficult to form.

For the gland shown in Fig. 5.1, the inner cylinder wall was formed in the early stage of deformation. When the outer wall and middle wall are stably extruded into the die, there was a splitting surface C (as shown in Fig. 5.3) of metla flow between the outer layer and the middle layer of the cylinder wall, and the flow velocity of the metal on both sides of the splitting surface is opposite. Assuming that the surface is located in the middle of the outer layer and the middle layer, then $S_2^f = 2331.6$ mm^2, $S_3^f = 5777.5$ mm^2 (the inner layer as the first layer), $S_2^p = 354.2$ mm^2, $S_3^p = 725.3$ mm^2, therefore $m_2 = 0.152$, $m_3 = 0.126$.

For the middle layer and the outer layer of the cylinder wall, there are $k_2 = 15.1$, $k_3 = 8.3$. Obviously, the second layer of the die cavity is filled in the end, that is, the control chamber should be designed at its end.

The annular control chamber is shown in Fig. 5.2b. The approximate relationship between the height of the annular control chamber h_c and the radius of the flange cavity r_0 can be obtained as

$$h_c = 0.082 \cdot r_0 \tag{5.3}$$

The outer radius of the annular control chamber r_c is

$$r_c = (1.1 \sim 1.15) \cdot r_0 \tag{5.4}$$

The volume of the corresponding annular control chamber is

$$V_k = 0.06 \cdot r_0^3 \quad (5.5)$$

The volume of the annular control chamber calculated according to Eq. (5.5) is approximately equal to 2 times the volume of the excess metal of the billet which is the difference between the upper deviation of the billet and the lower deviation of the forging volume.

When the last filled part of the cylindrical die cavity is determined, the inner diameter of the corresponding control chamber is also determined as $d_2 - 2t_2$ (d_2 is the outer diameter). According to the volume of the control chamber approximately equal to twice the volume of the excess metal of the billet, the height dimension of the cylindrical control chamber can be obtained

$$\Delta h_2 \approx \frac{V_k}{\pi t_2 \cdot (d_2 - t_2)} \quad (5.6)$$

5.2.3 Calculation of Forging Force

The force of flow control forming can be estimated using the following empirical equation:

$$p_b = cn\sigma_b F_m / F_b \quad (5.7)$$

where p_b is the unit pressure on the punch (MPa), σ_b is the tensile strength of the material at the forging temperature (MPa), F_m is the cross-sectional area of the billet (mm^2), and F_b is the cross-sectional area of the punch acting on the billet (mm^2), c is the hardening coefficient at the end of extrusion, and n is the restraint coefficient.

5.2.4 Finite Element Simulation of Forming Process

The model for the finite element simulation of the flow control forming process of the gland is shown in Fig. 5.4. This model was consistent with the actual backward extrusion process. The billet temperature was 420 °C, and the die temperature was 230 °C. The velocity of the punch was 20 mm/s. The shear friction factor was 0.3. The number of grids of the billet, punch and die were all 1000. When the grid distortion was large, the software will automatically remesh [3, 4].

Based on the numerical simulation results, the influence of the control chamber on the forging force was studied. The simulation results are shown in Fig. 5.5. It can be seen that there are three main stages during forming. Figure 5.5a shows the step when the outer cylindrical die cavity is filled. Figure 5.5b shows the step when the control chamber is just underfilled, and the forging force at this time is basically the

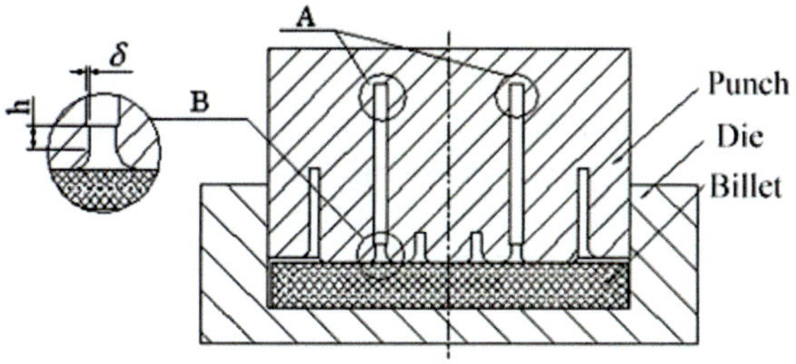

Fig. 5.4 Simulation model of forming process of gland

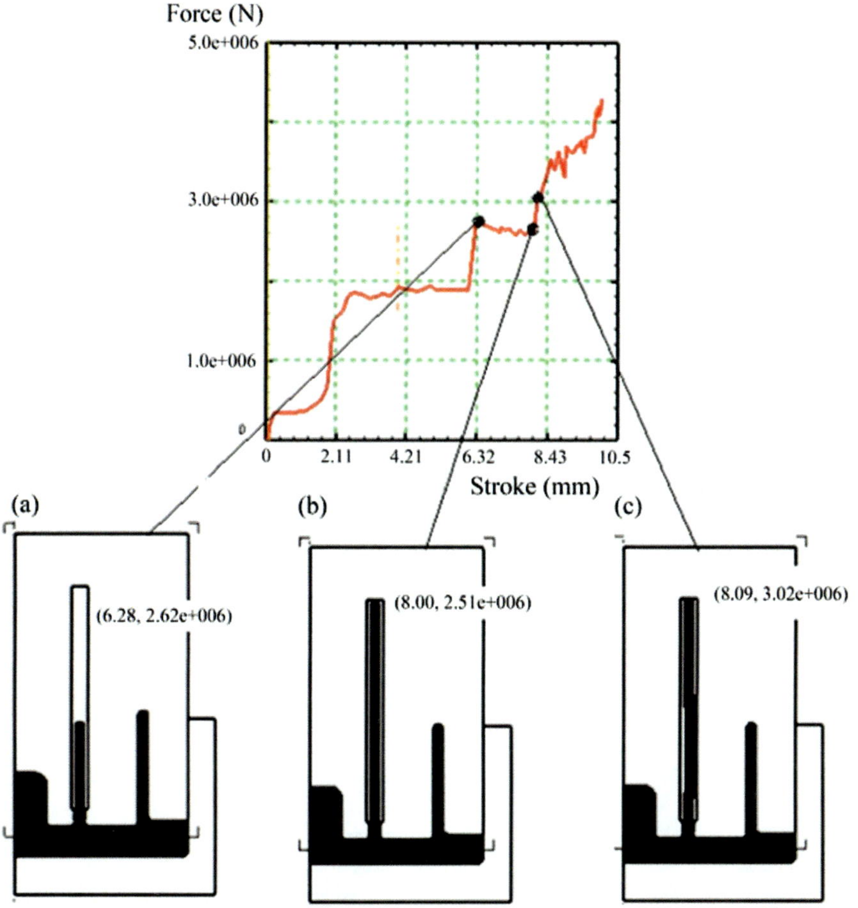

Fig. 5.5 Force–stroke curve of gland forming: **a** The outer cylinder is filled, **b** The control chamber is underfilled, **c** The control chamber is filled

same as Fig. 5.5a. After that, the excess metal flows into the control chamber. When the top of the forging contacts the bottom surface of the flow control chamber, the forging force increases sharply, as shown in Fig. 5.5c. By comparing of Fig. 5.5b, c, it can be found that during the movement of the punch by 0.09 mm, the forging force increases sharply from 2510 to 3020 kN with an increase of 20%. Through the above analysis, it is fully verified that the setting of the flow control chamber can reduce the forging force.

The following focuses on the analysis of metal flow and the velocity field during the backward extrusion process of the gland, as shown in Figs. 5.6 and 5.7. It can be seen that the forming process of the gland is roughly divided into three stages: the first stage is that the small round boss in the middle is completely formed at the initial stage of deformation; the second stage is that the outer cylinder is completely formed, the third stage is the inner layer cylinder is fully formed and the excess metal flows into the control chamber.

Due to the small ratio between height and diameter of the billet, the metal near the punch in the initial stage of extrusion process was in a steady flow state. With the continuous downward movement of the punch, the entire billet was in an unsteady flow state after the outer cavity was filled. It can be seen from Fig. 5.7 that in the stage when the punch is completely in contact with the billet and the outer cylinder is completely filled, there was a splitting surface between the inner cylinder and outer

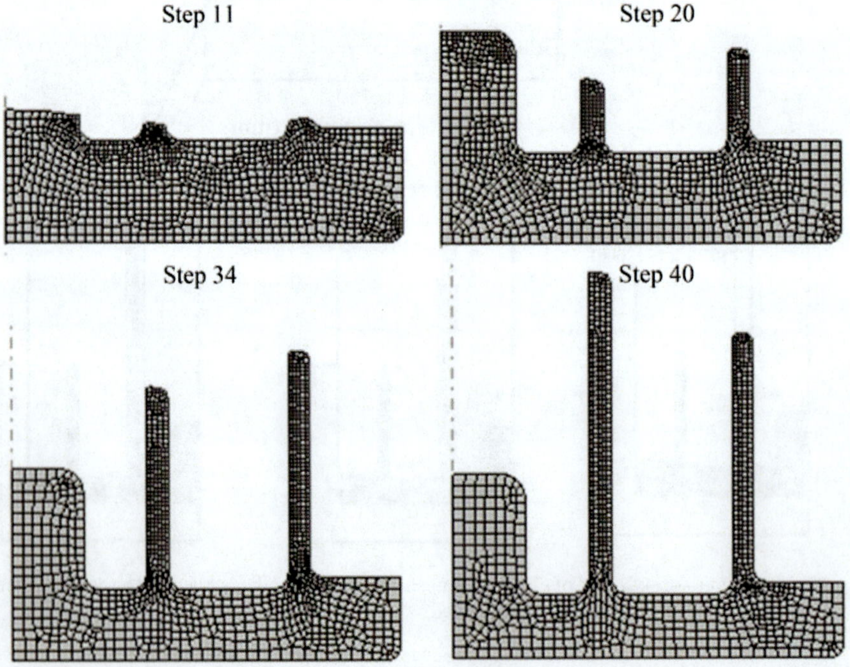

Fig. 5.6 The metal flow of the gland

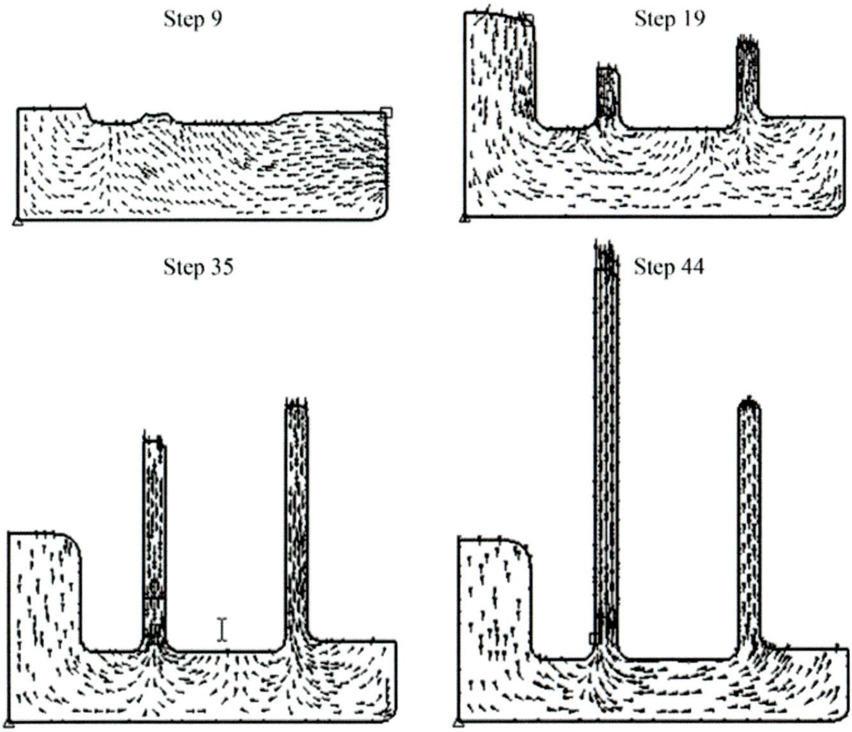

Fig. 5.7 Velocity distribution

cylinder. The flow direction of the metal on different side of the splitting surface is opposite, flowing to the inner and outer cylindrical cavity, respectively.

5.2.5 Die Design and Process Test

The schematic of the die structure of the gland forging is shown in Fig. 5.8. The punch was a combined structure which was composed of three parts 13, 14, and 15. The part 15 also acted as a stripper to remove the forging from the punch. The die was also a combined structure which was composed of three parts 4, 5, and 16. The part 4 also acted as an ejector which was responsible for ejecting the forging from the die. The part 17 fixed the part 16 on the lower template, and at the same time a certain compressive stress was formed on the part 16 to improve the bearing capacity of the die.

The die was installed on a Y28-400/400 double-acting hydraulic press which has inner slider and outer slider with a nominal force of 4000 kN, respectively. The inner slider and outer slider can be locked for use together, and the nominal force

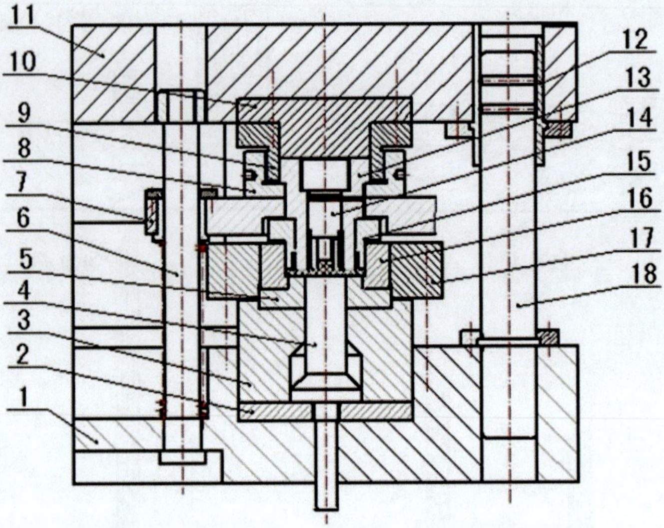

Fig. 5.8 Extrusion die for the gland. 1-Lower template. 2-Ejector backing plate. 3-Die backing plate. 4-Ejector. 5-Die 1. 6-Stripper tie rod. 7-Stripper plate. 8-Punch fixing screw. 9-Punch fixing ring. 10-Punch backing plate. 11-Upper template. 12-Guide sleeve. 13-Punch 1. 14-Punch 2. 15-Punch 3. 16-Die 2. 17-Die fixing ring. 18-Guide post

is 8000 kN. The nominal force of the ejection cylinder is 100 kN, and the force is transmitted to the ejector through the ejector rod.

The size of the cylindrical bars of the gland and housing were Φ102 mm × 12.65 mm and Φ86 mm × 18.7 mm, respectively. The bars were cut with a band saw, and then the two ends were flattened. The bars were heated by a box electric furnace. Before being loaded into the furnace, oil and other contaminants should be removed from the bars. When the electric furnace was preheated to 300 °C, the bars were put in. And then the furnace was heated to 420 °C and held for 1 h. The obtained gland and housing are shown in Fig. 5.9. Their dimensional accuracy and pressure resistance measured by the water burst experiment met the technical standards of airbags.

5.3 Closed Precision Forging of 4032 Aluminum Alloy Scroll with Back Pressure

5.3.1 Comparison of Forming Methods of Scrolls

Compared with the piston compressor, the scroll compressor has the characteristics of small starting torque, continuity of work, and higher pressure. The requirements of

Fig. 5.9 The formed gland and housing

machining accuracy of the main parts of the scroll, especially the shape and position tolerance of the scroll body, is very high. The perpendicularity of the end plane and the side wall of the scroll body should be controlled in micron level. Figure 5.10 shows the structure of the forging of scroll on a car air-conditioning compressor KC88.

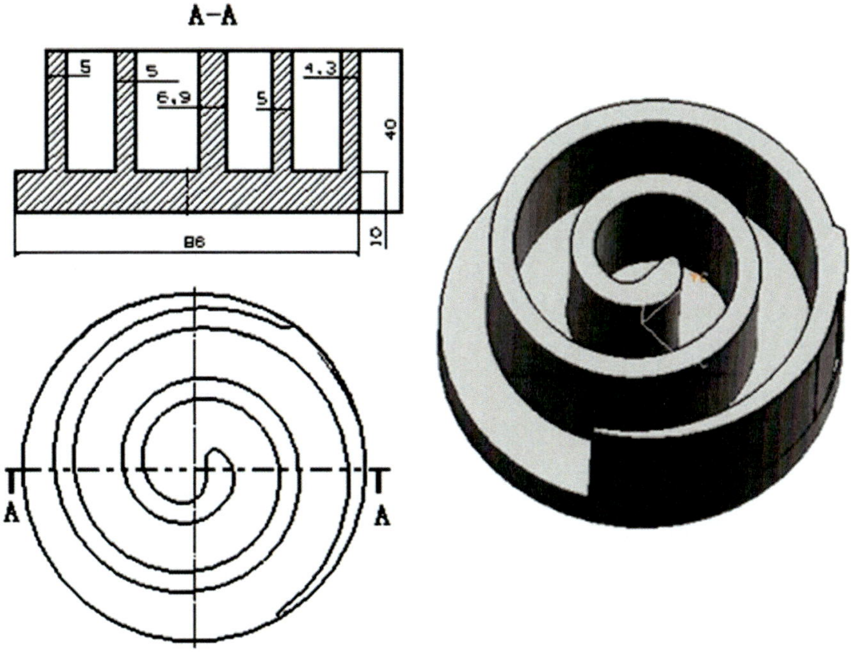

Fig. 5.10 Structural diagram of scroll

Currently, there are two main forming methods for scrolls, including squeeze casting and closed precision forging with back pressure. The former is realized by the squeeze casting die installed on the squeeze casting machine. It can obtain parts with ideal shape and size, but cannot obtain high-performance parts. The later is applying reverse pressure (ie back pressure) to the formed end surface of the scroll wall, so that the resistance of the part where the metal flows quickly is increased. The metal flow of this part is inhibited, thereby keep the height of the ends of the scroll flattening. This section will discuss the forming law and effect of back pressure on closed precision forging through numerical simulation and physical experiment [5, 6].

5.3.2 *Finite Element Simulation of Forming Process*

The finite element simulation model of forming process is shown in Fig. 5.11. The DFEORM-3D software was used to simulate the flow control forming process. The material of the parts was 4032 aluminum alloy. The velocity of the punch was 20 mm/s. The back pressure was 50 kN. The billet temperature was 470 °C, and the die temperature was 300 °C. The shear friction factor was set as 0.3.

In order to illustrate the flow control forming process with back pressure of the scroll, the forming process without back pressure was also simulated, as shown in Fig. 5.12. It can be seen that the conventional forward extrusion cannot correctly form the forging, the metal flow is extremely uneven, resulting in uneven end surface.

Figure 5.13 shows the simulation results of extrusion process with back pressure. It can be seen that the closed extrusion forming process with back pressure can ensure the flatness of the ends of the scroll on the one hand, and on the other hand, it can cause a strong three-dimensional compressive stress state and improve the plastic formability of 4032 aluminum alloy.

Fig. 5.11 The simulation model of flow control forming of the scroll

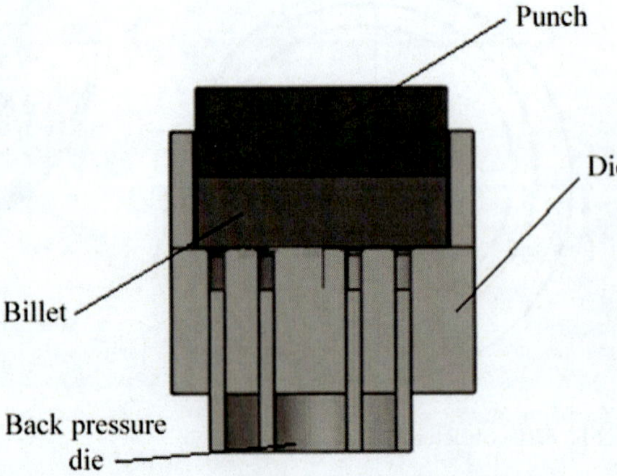

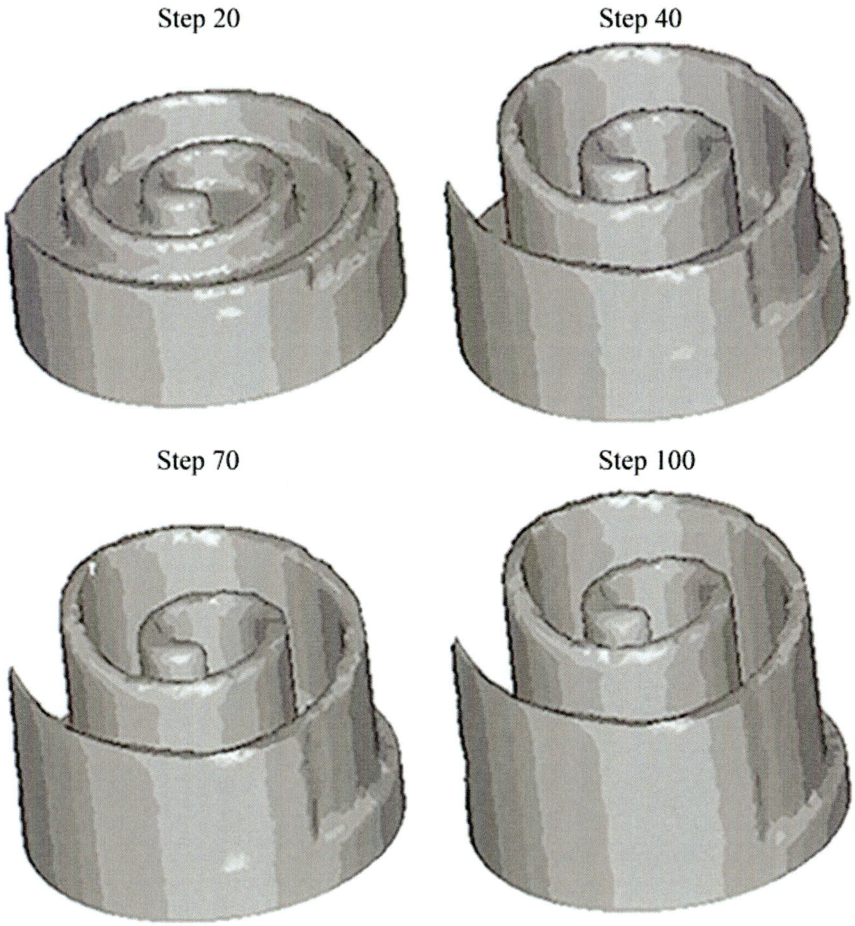

Fig. 5.12 Extrusion process without back pressure

5.3.3 Process Test

The structure of the designed extrusion die for scroll is shown in Fig. 5.14. It consisted of three parts: upper die, lower die, back pressure device. The upper die was composed of the punch, the punch fixing ring, the punch seat, the punch backing plate, the upper template and the upper die backing plate. The lower die was composed of the die, the pre-tightening ring, the scroll body, the die backing plate, the die seat, the lower backing plate and the lower template. The back pressure device was composed of the bush, the piston, the cylinder bottom plate, the sealing rings and the lower ejector rod. The upper die and the lower die were guided by the guide post and the guide sleeve.

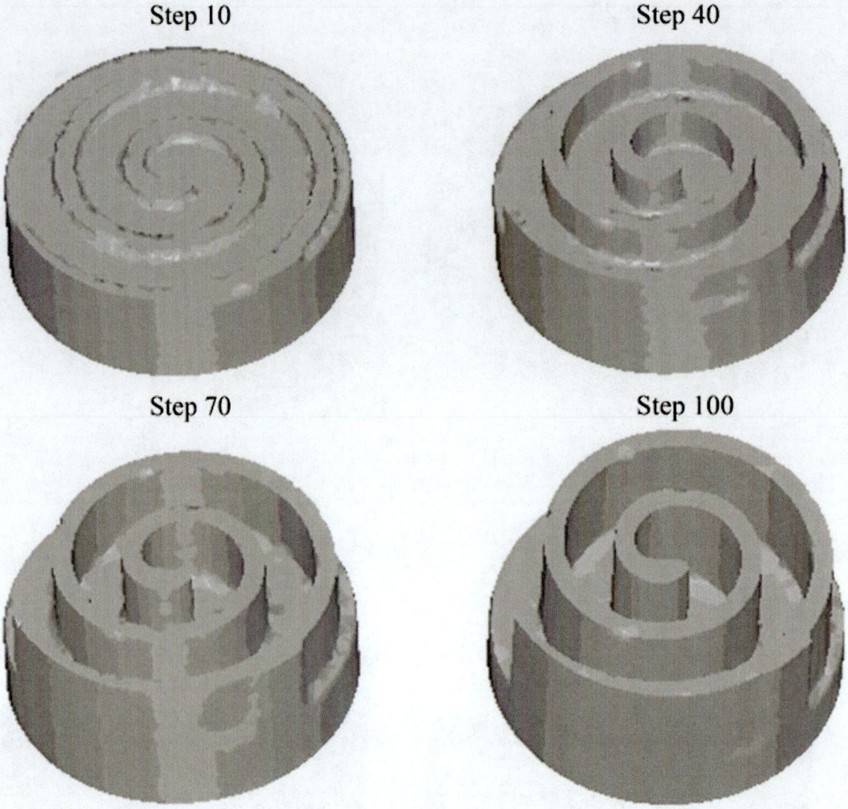

Fig. 5.13 Extrusion process with back pressure

During the extrusion of the scroll, the back pressure generated by the oil in the lower cavity of the oil cylinder piston exerted a reaction force on the billet through the lower ejector rod and the scroll body, forcing the deformed metal to the hard-to-fill zone. The magnitude of the back pressure was adjusted by the overflow valve. After extrusion, the upper die returned with the slider of press, and the lower chamber of the oil cylinder piston was pressurized with oil to move the lower ejector rod and the scroll body upwards, thereby the scroll was ejected from the die. It is worth noting that the die structure was extremely complex, and the machining accuracy and surface quality requirements were very high.

The precision forgings of scroll manufactured by the above-mentioned die are shown in Fig. 5.15.

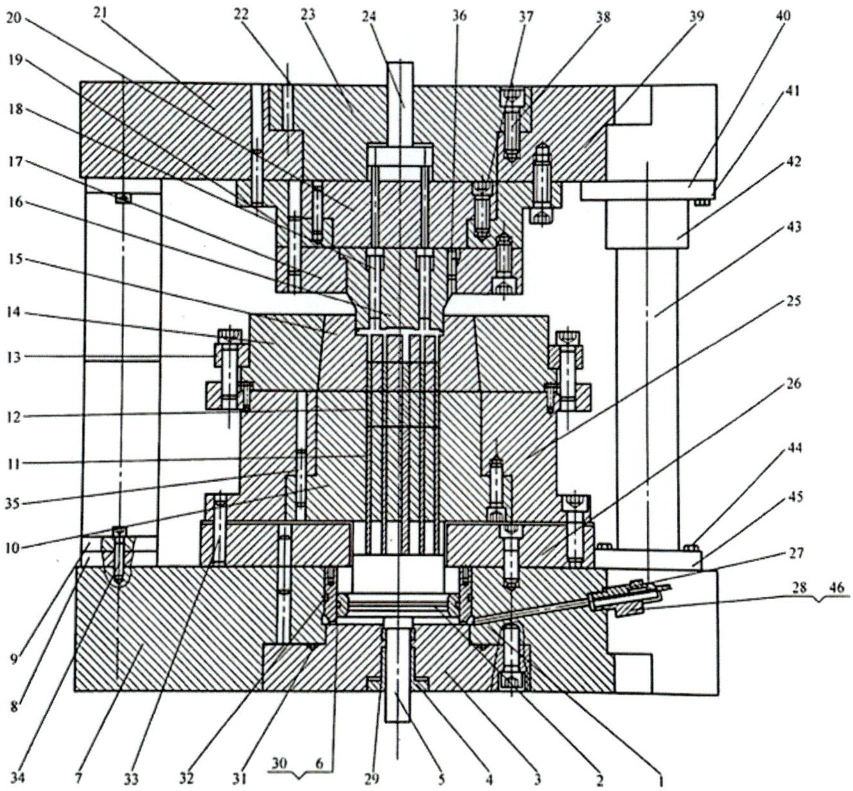

Fig. 5.14 Extrusion die of the scroll. 1-Bush. 2-Piston. 3-Cylinder bottom plate. 4-Guide sleeve of ejector rod. 5-Ejector rod. 6, 31, 32-Seal ring. 7-Lower template. 8-Cushion block. 9-Limit block. 10-Die backing plate. 11-Lower ejector rod. 12-Scroll body. 13-Pressing ring. 14-Pre-tightening ring. 15-Die. 16-Punch. 17-Punch fixing ring. 18-Upper ejector rod. 19-Punch seat. 20-Punch backing plate. 21-Upper template. 22-Screw hole. 23-Upper die backing plate. 24-Upper ejector. 25-Die seat. 26-Lower backing plate. 27, 28, 46-Hydraulic pipe joint assembly. 29-Guide ring of ejector rod. 30-Cylinder liner. 33, 35, 36, 41, 44-Pin. 34, 37, 38-Screw. 39-Upper template. 40, 45-Fixing ring. 42-Guide sleeve. 43-Guide post

5.4 Closed Precision Forging of 7075 Aluminum Alloy Tailstock

5.4.1 Process Analysis

The three-dimensional model of the tailstock forging is shown in Fig. 5.16. It can be deduced that there were two solutions for the precision forging of the part. One was to use ϕ74 mm at the lower end as the diameter of the billet. The other was to use ϕ84.5 mm at the upper end as the diameter of the billet. When the former was employed, the bottom of the lower end was formed by forward extrusion, and

Fig. 5.15 Precision forgings of the scroll

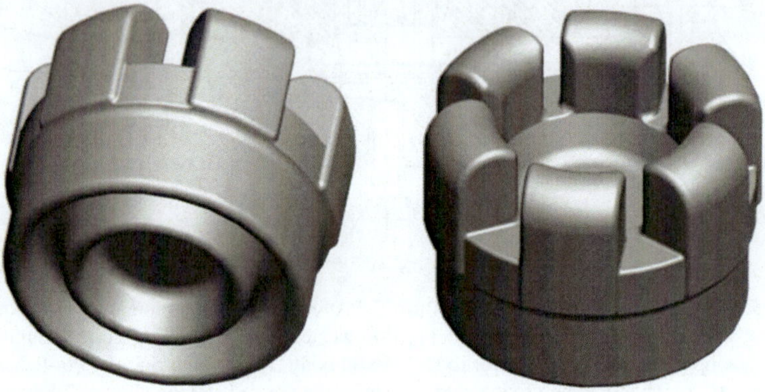

Fig. 5.16 Three-dimensional model of tailstock forging

the upper end was formed by upsetting and backward extrusion. When the later was employed, the lower end was formed by forward extrusion, and the upper end was formed by backward extrusion. Due to the poor plastic performance of 7075 aluminum alloy, when the former was employed, the upper end was in an open upsetting state, and the outer surface of the forging was in a state of tensile stress which may cause cracks.

5.4.2 Finite Element Simulation of Forming Process

The forming process of tailstock obtained by finite element simulation can be divided into three stages. The first stage was the backward extrusion stage. Because the two annular cavities at the bottom of the die were narrow and deep, the flow resistance of the metal at the bottom of the billet was large. The second stage was the forward and backward compound extrusion. When the six bosses with a rectangular cross section were extruded to a certain height, the flow resistance increased. While the metal at the top of billet continued to be extruded upwards, the metal at the bottom was extruded into the two annular cavities, that is, forward and backward compound extrusion. This stage is the main deformation stage. The third stage was the cavity filling stage. The annular cavity at the bottom of the cavity was the last filled part.

Considering that the outer wall was finally filled, the bottom of the outer cavity was deepened down by 3–4 mm as a relief cavity. At the end of forging, the effective contact area between the bottom of the forging and the die was reduced to 76.7% of the full contact area. Correspondingly, the forging force was reduced by about 1/4, which not only reduces the energy consumption of equipment, but also reduces the stress of the die.

5.4.3 Die Design

The forging die of the tailstock is shown in Fig. 5.17. The die was divided into two parts: upper die and lower die. The upper die was composed of the punching rod, the

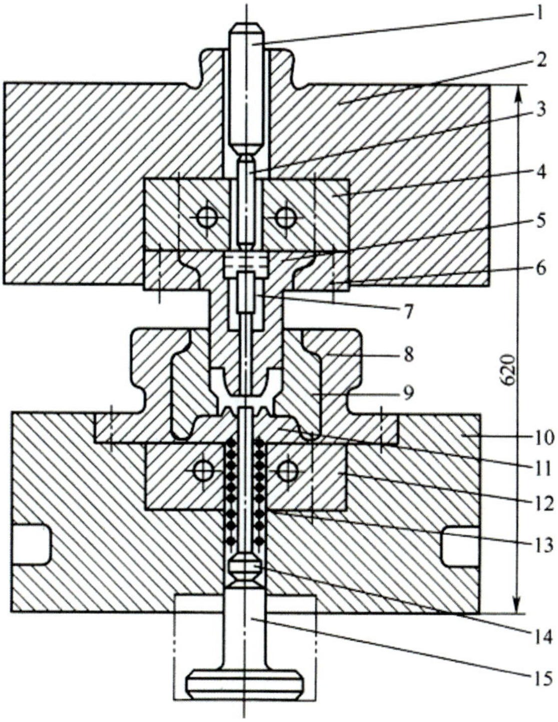

Fig. 5.17 Forging die of the tailstock. 1-Punching rod. 2-Upper template. 3-Pushing rod. 4-Upper backing plate. 5-Upper punch. 6-Punch fixing plate. 7-Upper ejector rod. 8-Die fixing ring. 9-Die. 10-Lower template. 11-Lower punch. 12-Lower backing plate. 13-Spring. 14-Lower ejector rod. 15-Ejector

upper template, the pushing rod, the upper backing plate, the upper punch, the punch fixing plate and the upper ejector rod. The lower die was composed of the die fixing ring, the die, the lower template, the lower punch, the lower backing plate, the spring, the lower ejector rod and the ejector. The die could be installed on a mechanical press or a screw press. Due to the change of the billet volume, longitudinal burrs might be formed in the annular gap between the upper punch and the die during forging. The forging might be stuck on the upper punch or in the die, so the upper ejector rod and the lower ejector rod were provided in the die at the same time.

In order to prevent the temperature of the billet from falling too fast during forging and affecting its forming performance, both the upper backing plate and the lower backing plate were provided with "U"-shaped electric heating tubes. During forging, the "U"-shaped electric heating tubes were used to preheat the forging die, and the preheating temperature was not less than 180 °C. Sodium disulfide plus engine oil or aqueous graphite was used as a lubricant. However, the two lubricants affected the surface quality of forgings, and it was better to use polymer lubricants.

5.5 Closed Precision Forging of 6061 Aluminum Alloy Wheels

5.5.1 The Forged Aluminum Alloy Wheels

Forging is one of the earlier forming processes for aluminum alloy wheels. Forged wheels have the advantages of high strength, good corrosion resistance, accurate size, and the good flow line consistent with the outline of forging. The strength, toughness and fatigue strength of forged wheels are significantly better than cast wheels. The elongation of the forged wheel is 12–17%, so it can absorb vibration well. Compared with cast wheels, forged wheels have a higher strength-to-weight ratio. In addition, the forged wheel has no pores on the surface, so it has a good surface treatment performance. The major disadvantage of forged wheels is that there are several production steps, and the production cost is much higher than that of casting.

At present, the main forging methods for producing aluminum alloy wheels are isothermal forging and closed precision forging [7]. With the advancement of technology, closed precision forging is most commonly used.

5.5.2 Closed Precision Forging with a Vertically Separable Die

1. Process design

The aluminum alloy wheel is shown in Fig. 5.18. The material of the part is 6061 aluminum alloy. It can be seen that the basic shape of the part is cup-shaped, and the

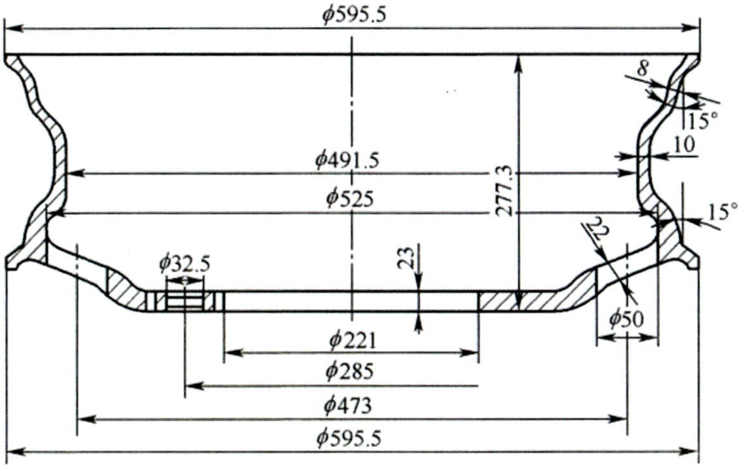

Fig. 5.18 Aluminum alloy wheel

wall thickness varies from 8 to 22 mm. The diameter of middle section is smaller than that of the upper and lower sections, which is similar to the double-drum shape that often appears in the upsetting step of the free forging.

According to the general design principles, the parting surface of the forging should be set at the top end with a large cross section (scheme 1), as shown in Fig. 5.19a. If the forging was designed according to this shape, it should have a certain forging slope from the bottom end to the parting surface. When the excess metal was removed by machining, the flow line of the part would be cut, reducing the strength and resistance to stress corrosion. If the machining allowance was added uniformly along the shape of the part (scheme 2), as shown in Fig. 5.19b, the material utilization can be greatly improved, and the flow line of the part was basically continuous after machining, however, there was a problem that the forging could not be extracted from the die due to the shape restriction.

From the comparison of economy and applicability, it could be concluded that if a die with a special structure was designed to solve the ejection problem, scheme 2 should be the better solution. Therefore, scheme 2 was employed to design the

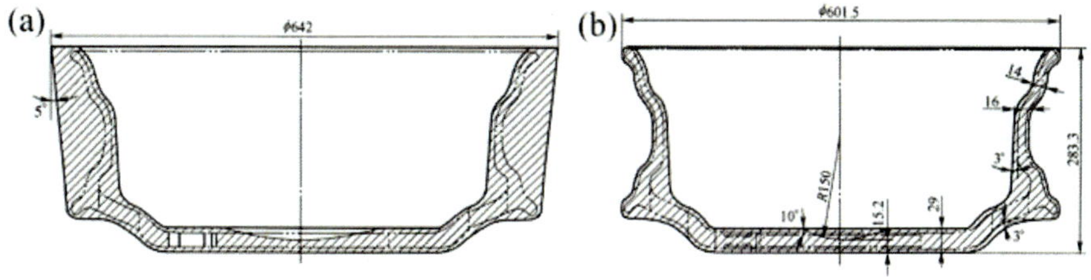

Fig. 5.19 Precision forging process. **a** Scheme 1, **b** Scheme 2

forging of wheel, that is, an even machining allowance of 3 mm and a small amount of necessary process remnants were added to the part.

2. Die design

Considering that the shape of the forging was a double-drum shape, the forging cannot be taken out when using an integral forging die. Therefore, the lower die was designed as a combined structure to ensure the forming of the forging and the extraction after forging. The lower die cavity was designed as a splitting insert die, and the insert box was used for positioning and guiding. The structure of the designed combined forging die is shown in Fig. 5.20.

When working, the die was installed on the PZS900 type 80,000 kN electric screw press. The heated billet was put into the die, and the press performed the first strike at 30–50% of the rated energy. After the billet was pre-filled, 80–95% of the rated energy was performed for the follow-up strike. Because the forging was a double-drum shape, the forming of the forging should be finished in one heating, otherwise the forging could not be put into the die again after the second heating due to the shape restriction. After the forging was formed, the inserts and the forging were ejected from the lower die together by the ejector, and the left and right inserts were automatically separated in the horizontal direction along the guide groove.

3. Process test

The preheating temperature of the die was 250–350 °C close to the final forging temperature (380 °C), which could effectively reduce the temperature drop of the

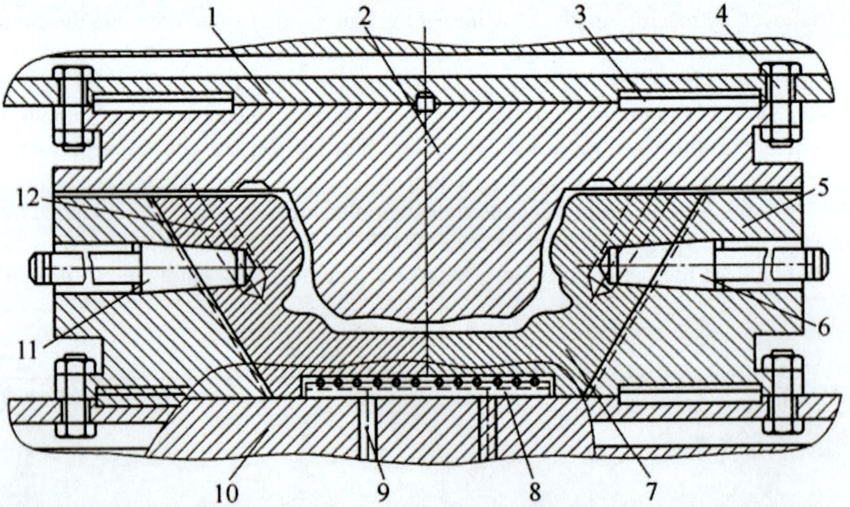

Fig. 5.20 The combined forging die. 1-Upper template. 2-Upper die. 3-Locating key. 4-Connecting bolt. 5-Insert box. 6-Right block pin. 7-Right insert. 8-Ejector guide plate. 9-Ejector rod. 10-Lower template. 11-Left block pin. 12-Left insert

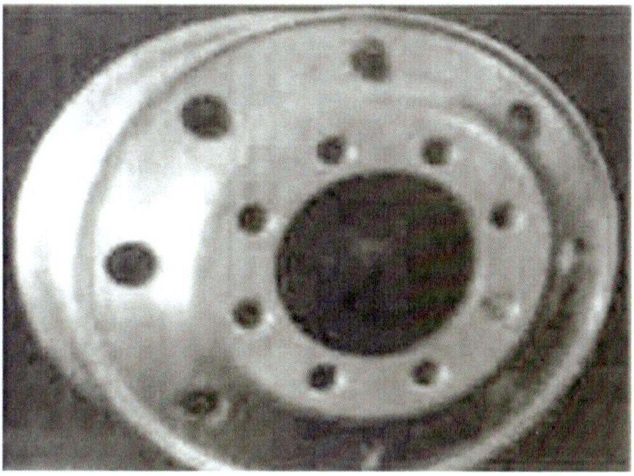

Fig. 5.21 The finished part of wheel

billet. The mechanical properties of the forging were qualified and fully met the requirements. The finished part is shown in Fig. 5.21.

5.5.3 *Closed Precision Forging with an Integral Die*

1. Closed precision forging process

In addition to closed precision forging with separable die, closed precision forging with integral die can also be used to form wheels. The closed precision forging process was completed by rotary forging, pre-forging, final forging, rim flaring, and spinning [8], as shown in Fig. 5.22.

The rotary forging is a kind of incremental forming with local linear contact, so the forging force is 1/5–1/20 of the die forging force. The microstructure of the billet obtained by rotary forging is uniform and fine, which improves the mechanical properties.

Due to the complex shape of the wheel, it was difficult to form directly in one step. The pre-forging process was added between the rotary forging and the final forging,

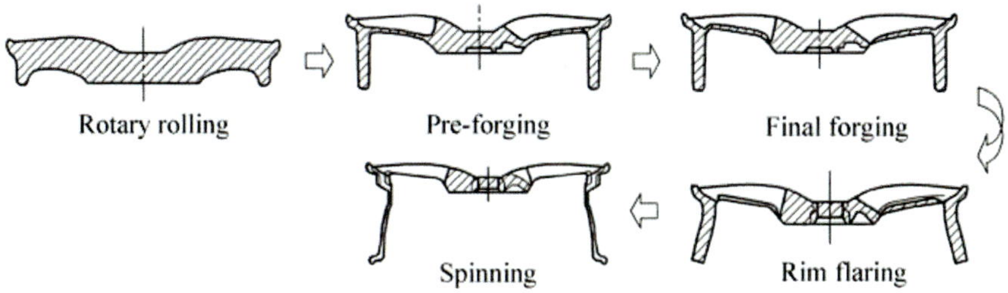

Fig. 5.22 Diagram of forming process of wheel

so that the pre-forged part was close to the shape of the final forging to ensure that the final forging was formed without defects. The closed precision forging was employed to realize pre-forming. Same as the principle of the pre-forging, closed precision forging was used for the forming of wheel spokes. Then, the rim was flared and spinned to obtain the required dimensions. Among them, the pre-forging was the most critical process in the entire forging of the wheel. The following is the die design for this process.

2. Die design of pre-forging

The design principle of pre-forging die is that the billet obtained by rotary forging flows uniformly and smoothly in the pre-forging die, and the pre-forged part can be deformed uniformly in the final forging die. The designed pre-forging die is shown in Fig. 5.23. In the closed precision forging, the cavity was closed in advance, and the billet was formed in the closed die cavity.

3. Process test

The diameter of the bar for process test was ϕ203 mm. During the test, defects such as folding, insufficient rim, deformation of the spokes, and folding in the inner wall appeared. The optimizations of the shape of the billet and the transition fillet of the key parts of the die were studied.

The positioning of the billet in the pre-forging die needed to be accurate. However, the error of the manipulator could not meet the precise positioning requirements, resulting in insufficient forming of the rim, as shown in Fig. 5.24a. Therefore, the rotary forging billet was modified to make it self-adaptively positioned in the pre-forging die (Fig. 5.25). Finally, the defect of insufficient rim was solved (Fig. 5.24b).

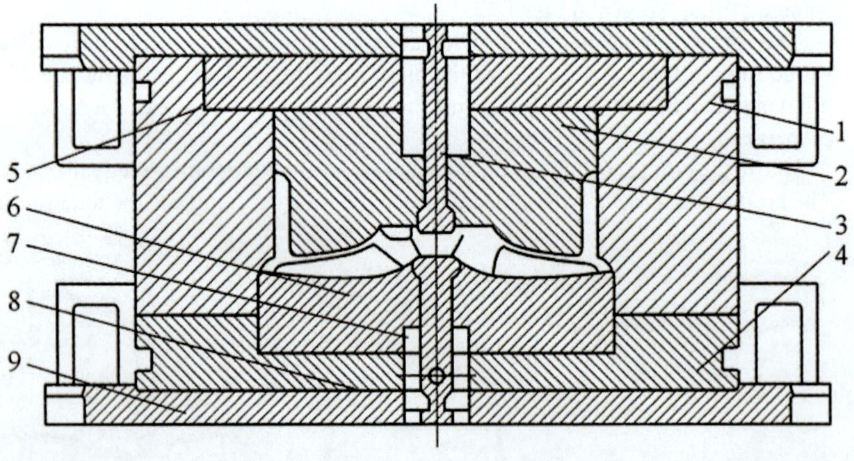

Fig. 5.23 Diagram of the pre-forging die. 1-Upper die cylinder. 2-Upper die. 3-Upper ejector. 4-Lower die seat. 5-Upper backing plate. 6-Lower die. 7-Lower ejector. 8-Stuck plate. 9-Lower template

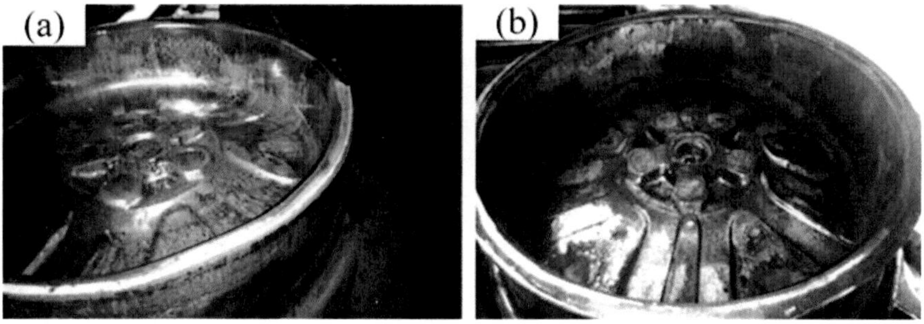

Fig. 5.24 The filling state of rim: **a** Insufficient filling, **b** Full filling

Fig. 5.25 Schematic of adaptive positioning of rotary forging billet

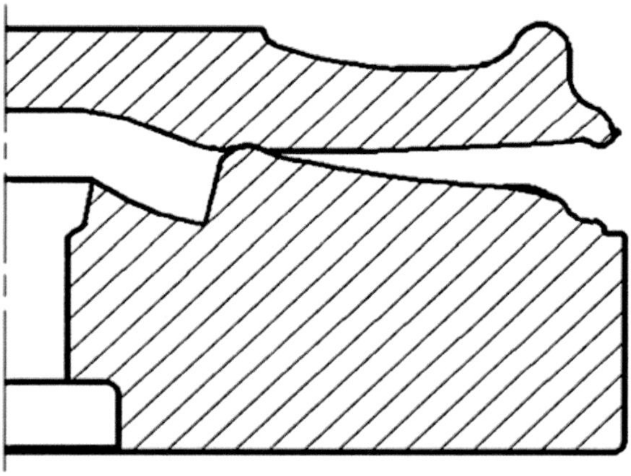

Fig. 5.26 The forged wheel

The transition fillet of the wheel in open die forging was usually 3–5 mm. However, in closed precision forging, due to the existence of metal backflow, the transition fillet should be larger than that of open die forging to avoid defects. In order to avoid the above defects, the transition fillet was optimized to 20 mm. The forged wheel without defects is shown in Fig. 5.26.

5.6 Isothermal Precision Forging of Aluminum Alloy

5.6.1 Isothermal Precision Forging

Isothermal precision forging refers to a forming method that keeps the die and billet temperature at the same or similar during the entire forging process with a slow forming velocity. When the material is forged at a higher temperature and a lower strain rate, sufficient recrystallization may occur, which can overcome the effect of work hardening.

The key to isothermal precision forging is that the billet deforms at a certain temperature or in a certain temperature range. For different materials, the optimal deformation temperature is different, so the control of temperature is very important. In order to obtain a constant temperature and a slow forming velocity, isothermal precision forging is generally carried out on a press equipped with a special heating device. The forgings obtained by isothermal precision forging have uniform microstructure, excellent mechanical properties, high material utilization, and good surface quality. Compared with conventional forging, isothermal precision forging overcomes the problems of die cooling, local overheating and uneven deformation. The microstructure and mechanical properties of isothermal forged parts have good uniformity and consistency. The plasticity of materials is significantly improved, and deformation resistance of the material is reduced. Due to the elimination of the influence of die chilling and strain hardening, not only the forging force becomes small, but also the forming process is simplified. Therefore, the large complex precision forgings are usually formed by this technology.

Figure 5.27 shows a rotor of the rocket engine. It can be seen that the shape of the part is complex, with thin and long blades whose length to thickness ratio is up to 10:1. It is not only difficult to be accurately formed by conventional forging methods, but also cannot be taken out from the die due to the twisted blades. The researchers used the isothermal precision forging technology to obtain good forgings at 420 °C using the die shown in Fig. 5.28 [9]. The flow line of the forgings was distributed completely consistent with the geometry of the blades without defects such as penetrating or turbulent flow. The microstructure was a completely recrystallized structure, and the grains were basically equiaxed. Tensile specimens of the blades were tested, and the tensile strength reached 413.5 MPa and the elongation was 29%.

Fig. 5.27 Aluminum alloy rotor [9]

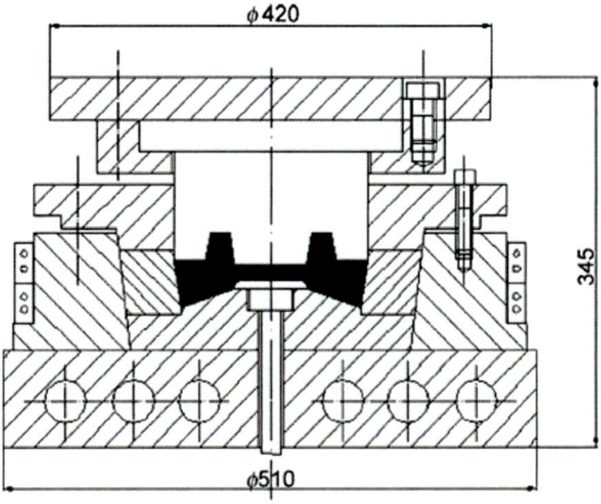

Fig. 5.28 Isothermal precision forging die of the rotor [9]

5.6.2 Isothermal Forging of 7075 Aluminum Alloy Piston

The traditional forming technology of the piston is casting. However, the casted part has coarse grains and poor mechanical properties, resulting in a short service life. The isothermal forging can obtain pistons with fine grains and high mechanical properties. The feasibility of the isothermal forging process of 7075 aluminum alloy piston and the grain size of the forgings were studied by the method of combining theoretical calculation, numerical simulation and experiment [10]. The research work was carried out in four aspects, including determining the forging process through theoretical calculation, using finite element simulation and experiment to verify the feasibility of the forging process, using the dynamic recrystallization evolution model of 7075 aluminum alloy to predict the change of grain size in the forming process, conducting the forging experiment to analyze the evolution of grain size.

1. Design of forging process

The aluminum alloy piston part is shown in Fig. 5.29. The skirt, ribs and pin holes of the piston were connected to form a closed irregular ring. The thickness of the skirt and ribs of the piston was thinner, and the contour at the ribs was an axisymmetric shape. As a whole, the part was a complex revolving parts.

The inner cavity of the piston skirt could be formed by backward extrusion. The contour at the two ribs were more complicated with through holes. If it was obtained by subsequent machining, the production efficiency and material utilization would become low, and the product performance would be damaged. Therefore, a multi-directional isothermal forging was employed to realize the forming of piston. The forging process was divided into two stages. The first stage was vertical backward extrusion to form the inner cavity of the piston. The second stage was horizontal extrusion to form the rib on both sides of the piston. The structural schematic of the die is shown in Fig. 5.30.

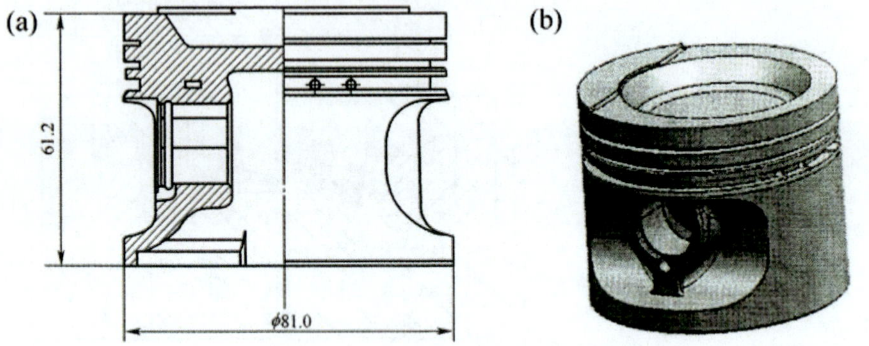

Fig. 5.29 The piston part: **a** Two-dimensional drawing, **b** Three-dimensional model

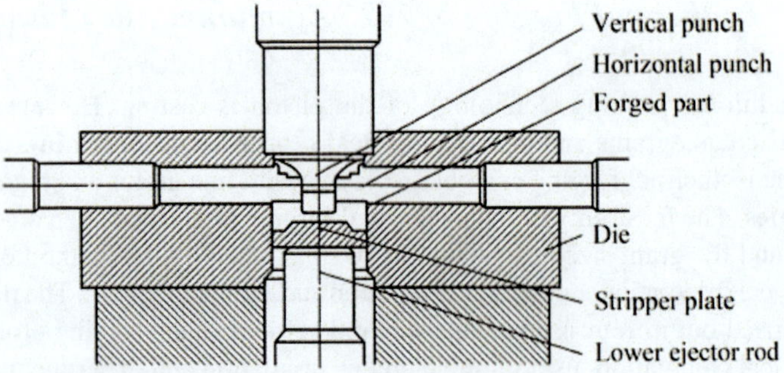

Fig. 5.30 Schematic of multi-directional forging die

In order to ensure that no defects such as cracks occurred during the forming process, the deformation degree of the extrusion were designed within the allowable deformation degree of the aluminum alloy. If the degree of deformation of one extrusion exceeds the allowable deformation degree of the material, multi-step extrusion or the pre-extrusion process should be considered. The allowable section reduction rate of aluminum alloy backward extrusion is 90–99%. The cross-sectional reduction rates of the piston longitudinal extrusion and transverse extrusion were 52.3% and 47.4%, respectively. Therefore, it can be determined that the extrusion process of the piston was one longitudinal extrusion and one transverse extrusion.

2. Numerical simulation of metal flow and grain evolution

In order to simplify the model, the die was set as rigid without considering the elastic deformation, the die temperature was nconstant during forging without considering the heat loss. The studied range of punch velocity was 0.125–1.25 mm/s. The finite element model is shown in Fig. 5.31.

The exponential model based on phenomenological theory was used to describe the evolution of average grain size during dynamic recrystallization [11].

$$d_{\mathrm{CDRX}} = a d_0^n \dot{\varepsilon}^m \varepsilon^p \exp \frac{Q}{RT} \tag{5.8}$$

where d_{CDRX} is the average grain size after dynamic recrystallization, d_0 is the initial grain size, $\dot{\varepsilon}$ is the strain rate, ε is the strain, T is the temperature, Q is the activation energy, and a, n, m, and p are the parameters.

A series of compression experiment was performed on 7075 aluminum alloy with an initial grain size of $d_0 = 40$ μm. Based on the experimental results, the fitting parameters were: $n = 0$, m $= -0.2$; $p = -0.3$; $Q = -11300$ J/mol; $a = 150$. The above model was integrated into the finite element software to predict the grain size during the isothermal forging process of piston. In the simulation, the temperature of the material and the die were both 380 °C.

Figure 5.32 shows the shape of the forging in different forming stages. At the 30th step, the metal was being extruded longitudinally to form the inner cavity of

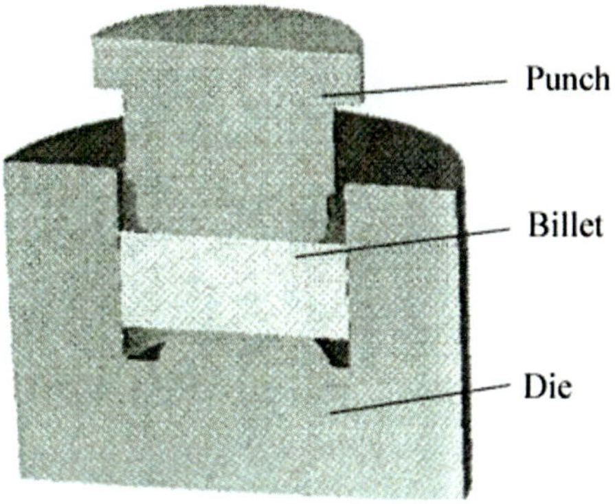

Fig. 5.31 The finite element model of the forging process of the piston

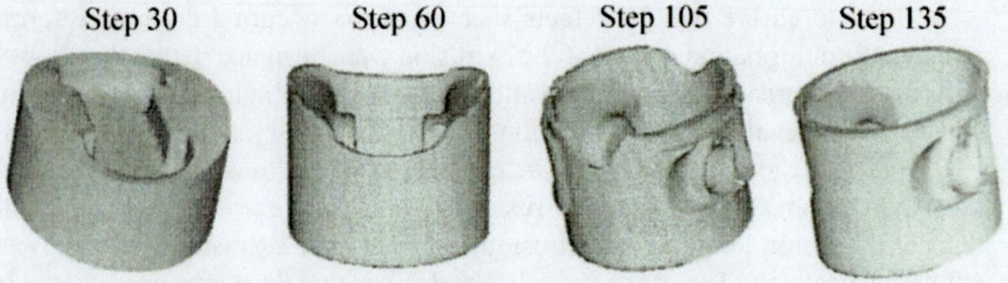

Fig. 5.32 The shape of the foring in different forming stages

the piston. At the 60th step, the longitudinal extrusion was completed and the inner cavity was formed. It can be seen that the ribs were formed faster than the thin side of the piston skirt during the longitudinal backward extrusion, and there was a larger gap in the skirt. At the 105th step, the pin holes and rib structure on both sides were extruded horizontally, while the skirt gap was gradually reduced. At the 135th step, the piston was completely formed. The results show that the metal flowed smoothly, and all parts of the piston were filled completely, and there were no defects. This indicated that it is feasible to use isothermal forging to form the piston.

Figure 5.33 shows the distribution of stress and strain at the end of forming. It can be seen from the stress distribution that the stress concentration was mainly located in the transition area of the punch fillet. At the edge of the rib and the ring surface outside the pin hole, there was a large stress concentration and the maximum stress was 129 MPa. But overall, the stress distribution was relatively uniform, about 50 MPa. It can be seen from the strain distribution that the strain at the round corners of the punch was larger which is similar to the stress distribution. The strain on the ring surface outside the pin hole and the edge of the rib was the largest, with a maximum strain of 14.8.

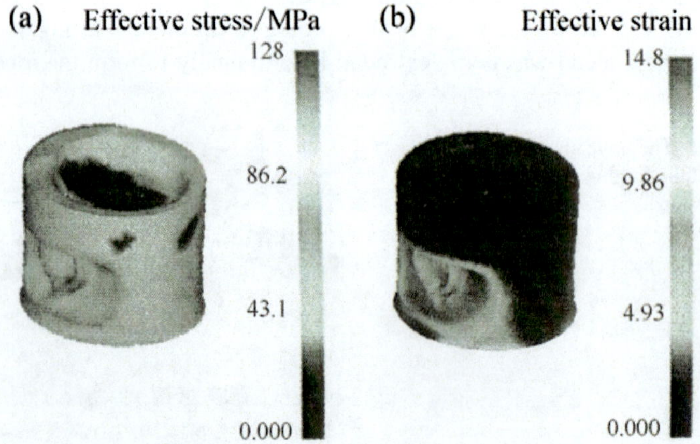

Fig. 5.33 The distribution of variables: **a** Stress, **b** Strain

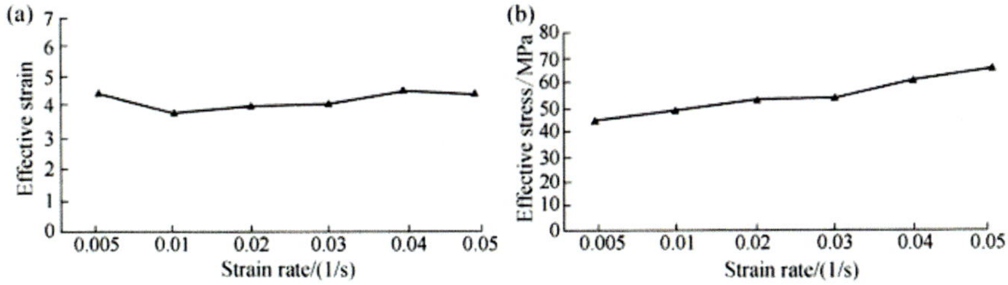

Fig. 5.34 The effect of strain rate on the strain (**a**) and stress (**b**)

In isothermal forging process, strain rate and forging temperature are the two main parameters. The effects of strain rate and forging temperature on strain and stress were studied by numerical simulation.

Figure 5.34 shows the effect of strain rate on the strain and stress. It can be seen that the effect of strain rate on strain was not significant, but the effect on stress was more obvious. This was because the increase of strain rate increases the dislocation density and the degree of work hardening, which leads to the stress of the metal in the deformation zone elevated.

Figure 5.35 shows the effect of forging temperature on the strain and stress. It can be seen that the forging temperature had a significant effect on the stress. As the forging temperature increased, the aluminum alloy underwent dynamic recovery, and dynamic recrystallization during the extrusion process, which led to an increase in the softening effect during the deformation process.

Figure 5.36 shows the predicted result of the grain size. It can be seen that the average grain size after forging was about 17.8 μm which was significantly reduced compared to the initial grain size. The grain size on both sides of the piston skirt and the bottom of the piston was about 28 μm, indicating that the refinement effect was not as obvious as that of other parts. The grain refinement was more obvious at the piston pin hole in which the grain size was refined to 13 μm. This was due to the greater plastic strain of the piston skirt during the longitudinal extrusion process. The energy generated by the severe deformation caused the dislocations to rearrange to form a substructure, and further evolved into low-angle grain boundaries and high-angle grain boundaries, thereby refining the grains. The metal at the pin hole

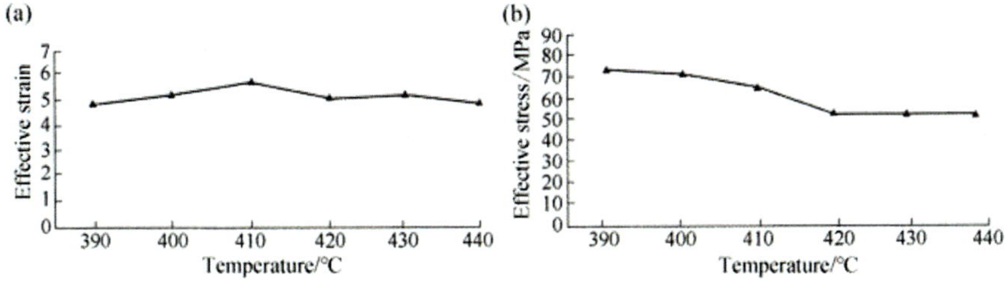

Fig. 5.35 The effect of forging temperature on the strain (**a**) and stress (**b**)

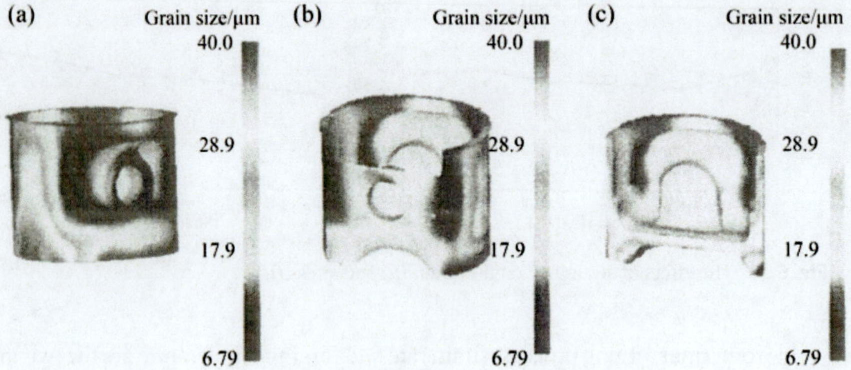

Fig. 5.36 Prediction results of grain size: **a** Overall view, **b** Half section view, **c** 1/3 section view

first underwent severe deformation together with the skirt during the longitudinal extrusion process, and then underwent severe plastic deformation again during the subsequent transverse extrusion process, resulting in the greatly refined grains.

3. Process test

The test material was 7075 aluminum alloy extruded bar of ϕ82 mm × 27.5 mm. The initial microstructure of the material is shown in Fig. 5.37. It can be seen that the grains were obviously elongated along the extrusion direction, and in cross section the grains were equiaxed. The initial grain size was 40 μm. After the billet underwent longitudinal extrusion and transverse extrusion on the hydraulic press, the aluminum alloy piston part was obtained, as shown in Fig. 5.38.

The metallographic samples were taken from the side of the piston skirt and the piston pin hole, respectively, as shown in Fig. 5.39. The microstructure of the side of the piston skirt was obviously finer than the initial microstructure. The grain size at this place was 26 μm. Compared with the piston skirt, the microstructure of the piston pin hole was more refined, and the grain size was 14 μm. The prediction

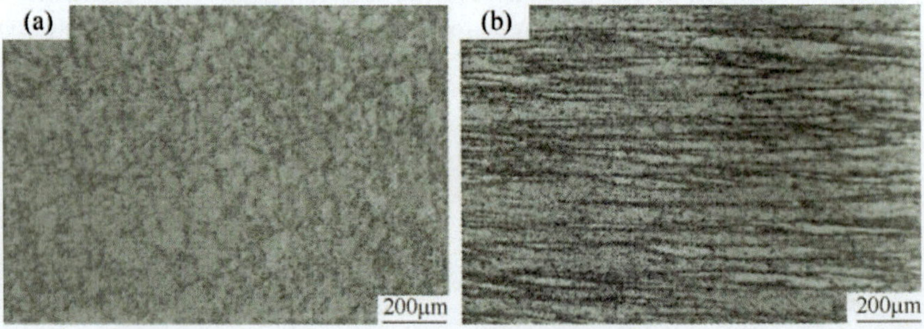

Fig. 5.37 Initial microstructure of 7075 aluminum alloy bar: **a** Cross section, **b** Longitudinal section

Fig. 5.38 The forged piston part

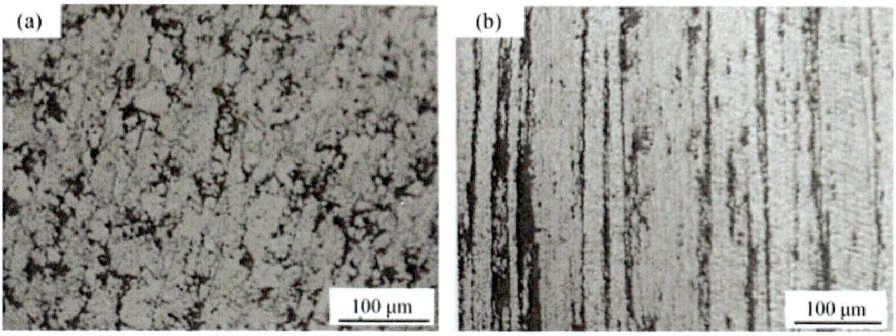

Fig. 5.39 Microstructure of different parts of the piston forging: **a** Skirt. **b** Pin hole

results were in good agreement with the experimental results, which provided useful guidance for controlling the microstructure of the forging.

5.7 Cold Precision Forging of 2024 Aluminum Alloy Driving Wheel

5.7.1 Process Analysis

The shape and dimensions of driving wheel are shown in Fig. 5.40. The material of this part was 2024 aluminum alloy. There were 40 trapezoidal teeth distributed

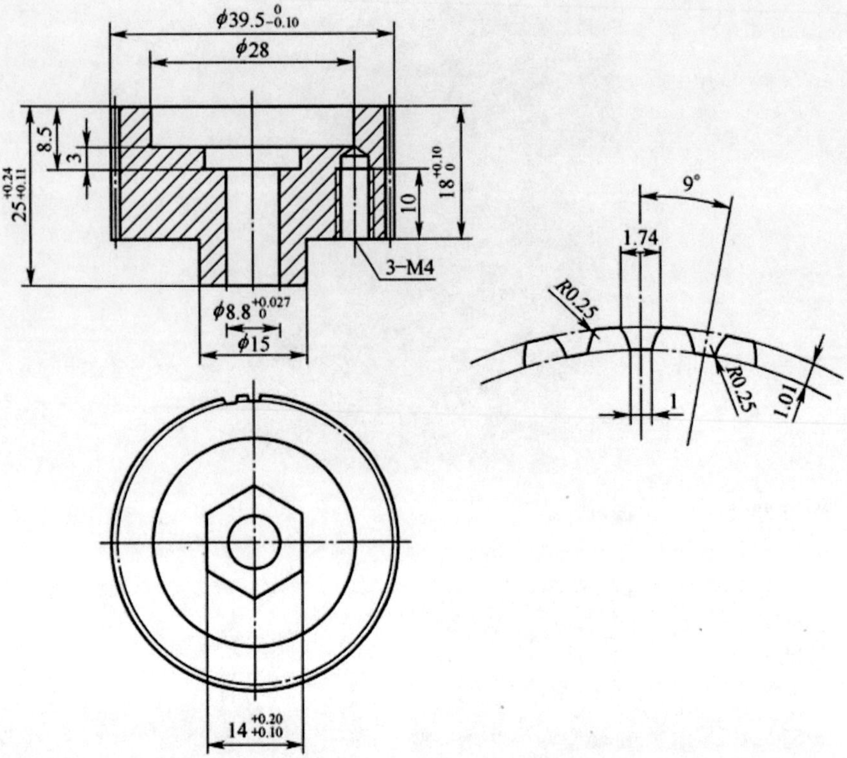

Fig. 5.40 Drawing of driving wheel

on the outer circumference to transmit torque. The bottom of the inner hole at the upper end of the part was an inner hexagon counterbore, which was used to place fastening screws. The trapezoidal teeth and hexagon counterbore could be formed by cold precision forging due to low cost compared with general machining.

Due to the high content of copper and magnesium, a large amount of compounds are formed in 2024 aluminum alloy. The compounds not only increase the strength of the material, but also greatly reduce the plasticity of the material. Therefore, the 2024 aluminum alloy in the supply state had high strength, large deformation resistance, and poor plasticity, making it difficult to perform cold precision foring with a large degree of deformation. Through sufficient annealing and softening treatment, the plasticity of 2024 aluminum alloy was significantly improved, the deformation resistance was also greatly reduced. With effective surface and lubrication treatment, it was feasible for this part to be formed by cold precision forging.

The key to the cold precision forging of this part was the forming of the trapezoidal teeth. The dimensional accuracy of teeth was IT11. Generally, the dimensional accuracy of cold precision forging was IT7–8, so cold precision forging could meet the dimensional accuracy requirements of this part.

5.7.2 Process Design

(1) Design of the forged part

According to the characteristics of the driving wheel and the cold precision forging process, the trapezoidal teeth, ϕ28 mm recessed holes and the hexagonal counterbore could be directly formed, and ϕ15 mm boss on the lower end could also be extruded. A certain amount of machining allowance was added on the outer diameter and end surface of the ϕ15 mm shaft, and the step surface of the ϕ15 mm shaft and the outer gear. Due to the small diameter of the ϕ8.8 mm inner hole and the requirement of concentricity with the trapezoidal teeth, it was difficult to ensure the accuracy, so the inner hole was obtained by subsequent machining. The upper end of the part was the free end, so a certain machining allowance was added. The drawing of the cold forging is shown in Fig. 5.41.

(2) Cold precision forging process

Taking into account the projection profile of trapezoidal teeth, a billet with the same diameter as the tooth root was used. During forging, a part of the metal flows upward and at the same time flows outward in the radial direction to form trapezoidal teeth, and the other part of the metal flows downward to form a boss. During the process test, it was found that the top of the trapezoidal teeth near the lower end was slightly unformed. In order to ensure the dimensional accuracy of the teeth, a precision sizing process was added. After sizing, the tooth profile was full and smooth, which met the requirements of the parts.

(3) Allowable degree of deformation

In cold forging, the excessive degree of deformation makes the deformation resistance increase sharply, resulting in a decrease in the service life of the die. Therefore, the

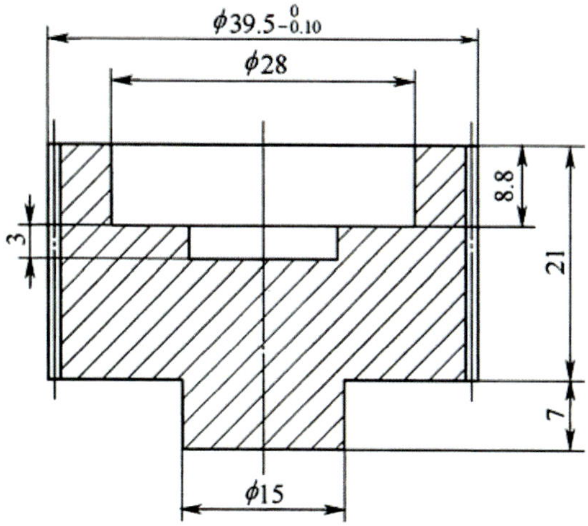

Fig. 5.41 Drawing of the cold forging of driving wheel

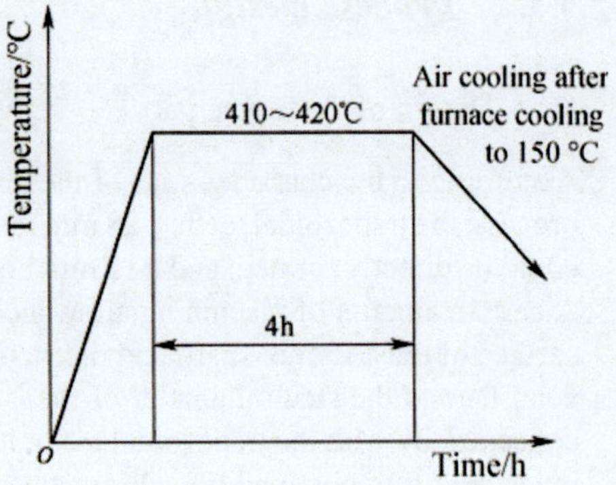

Fig. 5.42 Annealing treatment of the 2024 aluminum alloy

degree of deformation of the material during forging should be strictly controlled. The degree of deformation of cold precision is generally expressed by the reduction of area. When the ϕ15 mm boss was extruded, its reduction of area was 82%, and the reduction of backward extrusion was 55.7%. Generally, the allowable reduction of 2024 aluminum alloy is 90% and 75%, respectively. Obviously, the reduction of area of this part met the above requirements.

(4) Billet softening and surface lubrication

The annealing treatment was required for cold precision forging of 2024 aluminum alloy. The specification of softening annealing treatment is shown in Fig. 5.42. The hardness of the 2024 aluminum alloy billet after annealing was 50HB-60HB.

In the cold forging of driving wheel, it was necessary to maintain a good lubrication state between the billet and the die cavity. Due to the high unit press, lubricants were easily squeezed out and does not provide good lubrication. Therefore, surface treatment should be carried out before lubrication treatment. The surface treatment employed oxidation treatment which formed a porous oxide film on the surface of the billet. During forging, the film was deformed along with the billet and adhere closely to the surface of the billet to form a lubricant layer. The lubricant layer continuously released lubricant during forging to ensure that the deformed material and the die cavity were always in a good lubrication state. The oxidation treatment solution was sodium hydroxide (40–60 g/L) of 50–70 °C, the treatment time was 1–3 min. The industrial vegetable oil was used for the lubrication treatment of the cold precision forging of driving wheel.

5.7.3 *Die Design*

The partial structure diagram of the forging die is shown in Fig. 5.43. A three-layer combined structure was employed. The trapezoidal teeth were manufactured by wire cutting. Both the inner die and the die insert were made of Cr12MoV steel with the heat treatment hardness of 61–63HRC. The material of the middle reinforcement

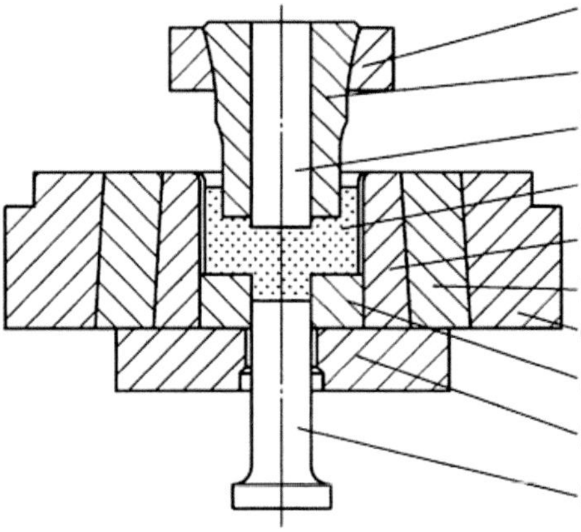

Fig. 5.43 Schematic of the partial structure of the forging die

ring was 35CrMoA steel with the heat treatment hardness of 42–44HRC. The heat treatment hardness of the outer reinforcement ring was 38–40HRC.

The punch alos employed a combined structure. There should be enough interference between the hexagonal mandrel and the hole of the punch to prevent the mandrel from loosening when the punch returned after the forging.

References

1. Xia JC, Hu GA, Wang XY, Cheng JW (2004) Flow control forming process analysis and forging force calculation of multilayer cup cylindrical parts. China Mech Eng 15(1):91–93
2. Xia JC, Hu GA, Wang XY, Cheng JW (2004) Study on flow control precision forming technology of airbag parts in car. Forging Stamp Technol 29(1):1–3
3. Cheng JW (2006) Study on the key technology of precision die forging for high strength Al-alloy parts in car. Huazhong University of Science and Technology, Wuhan
4. Zhang YR, Xia JC, Cheng JW, Hu GA (2007) Analysis on hot extrusion process and FEA of airbag part in car. J Plast Eng 14(2):73–76
5. Zhang YR (2007) Study on the hot extrusion forming and FEM simulation for multi-layer thin walls columnar parts. Huazhong University of Science and Technology, Wuhan
6. Deng L, Dai WL, Wang XY, Jin JS, Li JJ (2018) Metal flow controlled by back pressure in the forming process of rib-web parts. Int J Adv Manufact Technol 97:1663–1672
7. Zhou XH, Jia QX (2004) Precision die design and improvement for large-scale double-drum shape aluminum hub. Forging Stamp Technol 29(1):70–73
8. Ma ZY (2015) Study on closed-die forging technology of forging aluminum alloy wheel. Forging Stamping Technol 40(8):1–4
9. Shan DB, Liu F, Xu WC, Lv Y (2005) Experimental study on process of precision forging of an aluminium-alloy rotor. J Mater Process Technol 170(1–2):412–415
10. Zhou R, Sha B, Chen WL (2013) Investigation of Al-alloy complex part on isothermal extrusion process. In: Jinan: proceedings of the 5th national symposium on precision forging, pp 166–169
11. Yang D, Chen WL, Zhou R (2013) Microstructure grain size prediction of 7075 aluminum alloy prepared by piston isothermal extrusion. In: Jinan: proceedings of the 5th national symposium on precision forging, pp 161–165

Chapter 6
Combined Casting-Forging Process for Branch-Shaped Parts

6.1 Overview

Typical branch-shaped forgings such as aluminum alloy steering knuckles and control arms have complex structures. If the traditional die forging processes are employed to produce the forgings from the wrought aluminum alloy bar, although high strength and good quality of the forgings can be obtained, there will be problems such as multi-step billet-making and multiple heating, which will not only cause long production cycle and low efficiency, but also lead to low material utilization and high energy consumption. If the casting processes are employed to form the part, although it has the advantages of good fluidity and easy forming, there will be defects such as shrinkage cavities, porosities and coarse grains in the casting, causing the mechanical properties to fail to meet the requirements of use [1].

In recent years, the casting process and the forging process are combined to form a part to solve the above technical problems, that is, combined casting-forging process. Researchers have carried out the following two aspects of research on the combined casting-forging process: (1) the squeeze casting is used to obtain the pre-formed part, and then the semi-closed or closed precision forging is used for final forming; (2) the cast bar or cast block is taken as the billet of closed precision forging. The key to this process lies in the forging process which eliminates the shrinkage cavities and porosities that generate during the casting process and improves the mechanical properties of the part.

6.2 Precision Forming Process of Swash Plate

Figure 6.1 shows the structure of the swash plate type air-conditioning compressor. It can be seen that the key moving parts are the swash plate and the piston. These parts perform high-speed reciprocating motion during work. Therefore, good strength and

L. Deng et al., *Precision Forging Technology and Equipment for Aluminum Alloy*, Springer Series in Advanced Manufacturing,
https://doi.org/10.1007/978-981-19-1828-5_6

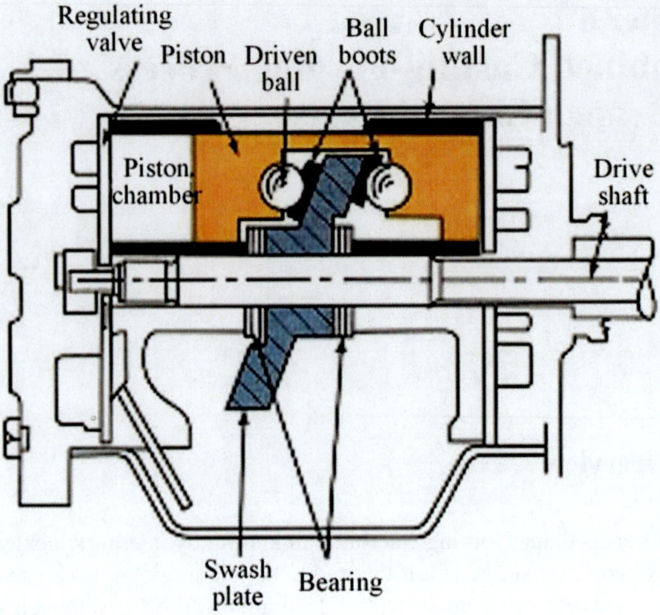

Fig. 6.1 Structure diagram of air-conditioning compressor

wear resistance are required for the two parts. Generally, the material of the swash plate is a high-silicon aluminum alloy A390 with a silicon content of more than 17%, and the material of the piston is a medium silicon aluminum alloy with a silicon content of 9.5% ~ 11.0%, such as 4032 aluminum alloy.

Medium and high silicon aluminum alloys have poor plasticity in solid state, but have good flow properties in liquid state. Therefore, the traditional production method of such parts is die-casting. However, there are shrinkage cavities and porosities inside the die-cast parts, which not only reduces the strength of the parts, but also affects the air tightness during work. In recent years, with the continuous improvement of the performance of air-conditioning compressors, die-cast parts can no longer meet the requirements of their use. The combined casting-forging process was proposed to manufacture higher quality parts [2–5].

6.2.1 Squeeze Casting Process

Figure 6.2 shows the two-dimensional sectional view of the swash plate casting. The center of the part was a $\Phi 39 \times 30$ cylinder, and the outside of the cylinder was a $\Phi 79$ swash plate. The angle of intersection between the swash plate and the axis was 68.5°. As mentioned earlier, the material of the swash plate is the high silicon aluminum

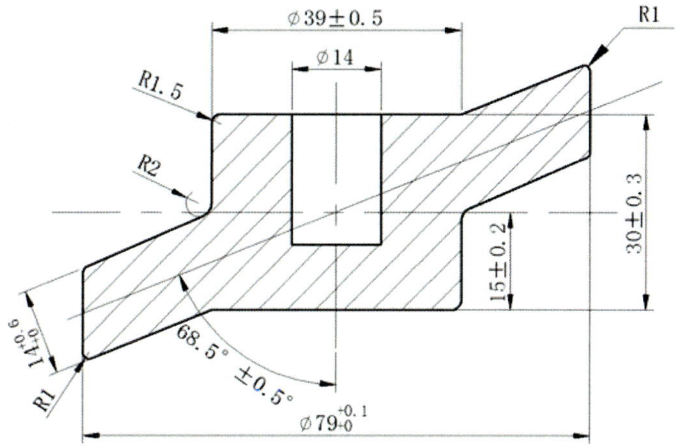

Fig. 6.2 The drawing of swash plate casting

alloy A390. The following four types of traditional squeeze casting process were designed to study the forming law [6].

1. Single-Punch Squeeze Casting

Figure 6.3 shows a schematic of the single-punch squeeze casting process. The whole process was divided into 5 steps. In the first step, the aluminum alloy melt was poured into the fixed cylinder (the cavity was composed of the top surface of the lower die at the lower limit position and the fixed cylinder wall), as shown in Fig. 6.3a. In the second step, the upper die moved downwards under the drive of the press slider, and closed with the fixed cylinder to form a cavity, as shown in Fig. 6.3b. In the third step, the lower die squeezed upwards under the push of the piston of the lower cylinder, so that the aluminum alloy melt filled the entire closed die cavity until the aluminum alloy melt was solidified under pressure to form a swash plate casting, as shown in Fig. 6.3c. In the fourth step, the upper die returned to the initial position, as shown in

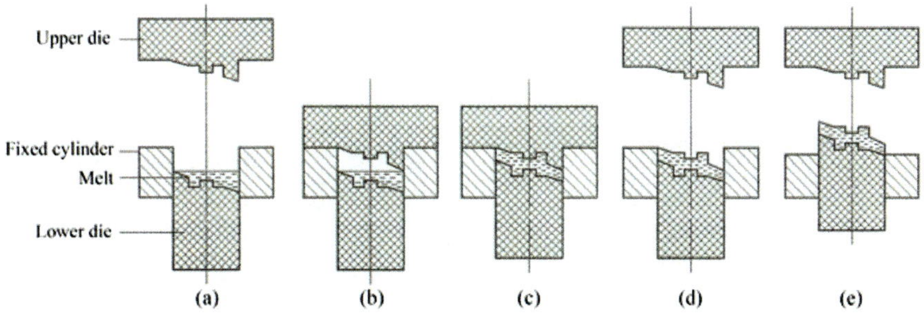

Fig. 6.3 Schematic of single-punch squeeze casting process: **a** Pouring, **b** Clamping, **c** Squeezing of lower die, **d** Returning of upper die, **e** Ejection of the part

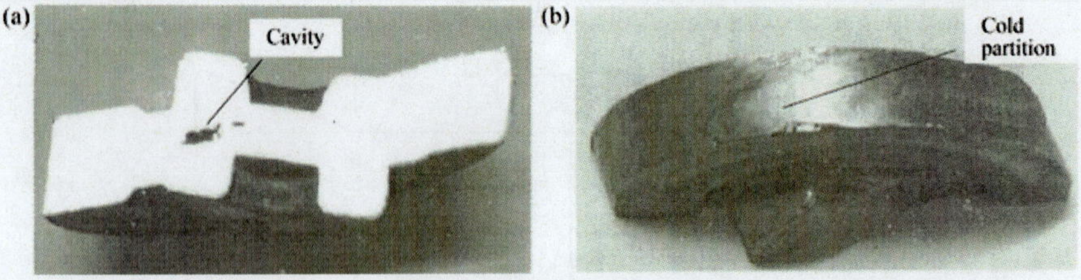

Fig. 6.4 Defects of single-punch squeeze casting: **a** Shrinkage cavity, **b** Cold partition

Fig. 6.3d. In the fifth step, the lower die pushed the casting out of the fixed cylinder under the action of the upward movement of the lower cylinder piston, as shown in Fig. 6.3e.

The castings formed by this process had defects such as shrinkage cavities and cold partition, as shown in Fig. 6.4. The generation of shrinkage cavities is due to the fact that the thin part of the edge of the swash plate solidified first, which hindered the transmission of the extrusion force, resulting in the solidification of the thick part in the middle could not be fed. Meanwhile, because the aluminum alloy melt stayed in the die cavity for a long time after it was poured into the die cavity, a layer of cold shell was formed on the wall. This cold shell could not be fused with the later solidified aluminum alloy, resulting in the formation of cold partition. The defects of the parts formed by this process were mainly caused by the die structure and could not be eliminated by adjusting the squeezing process parameters.

2. Double-Punch Squeeze Casting

The process of double-punch squeeze casting was the same as that shown in Fig. 6.3. The process test showed that the shrinkage cavity in the middle thick part was directly related to the squeeze depth of the upper die, as shown in Fig. 6.5. When the depth of the upper die was small, large shrinkage cavities appeared in the center. When the squeeze depth was 10 mm, there was still shrinkage cavities in the section. However, when the depth of squeezing increased to 20 mm, no defects were generated in the part.

The squeeze depth of the upper die was determined by the start time of the upper die. The longer the initial squeeze time, the more solidification of the alloy liquid, the

Fig. 6.5 The effect of the squeeze depth of the upper die on the shrinkage cavity: **a** Depth of 8 mm, **b** Depth of 15 mm, **c** Depth of 20 mm

more difficult it was for the upper die to squeeze into the material, and the weaker the feeding effect. Therefore, it was necessary to control the timing of squeezing with the upper die. Moreover, the upper die should be withdrawn at an appropriate time. If it was drawn out too early, the cavities would retract, which affected the feeding effect. In addition, if the holding time was too long, the upper die would stick to the casting, causing difficulty in demolding. The test results showed that when the upper die started to squeeze after the melt just filled the die cavity and immediately withdrawn after pressure holding for 5 to 7 s, the casting without defects could be obtained.

3. Indirect Squeeze Casting

The die structure of this scheme consisted of three parts: upper die, middle die, and lower die. A storage chamber was designed in the lower die. The indirect squeeze casting process is shown in Fig. 6.6. The casting is shown in Fig. 6.7.

In the squeeze casting process, after the liquid metal was poured into the storage chamber, a cold shell was quickly formed on the surface of the chamber. When the aluminum alloy melt was filling to the die cavity, the outer shell and coating were blocked out of the sprue. The aluminum alloy melt with higher temperature passed through the sprue and filled the cavity in a short time to avoid the cold partition. The remaining aluminum alloy melt with lower temperature and higher impurity content solidifies in the chamber formed a sinter cake. It can be seen from Fig. 6.7 that the surface of the casting was smooth and clean without cold partition. However, the handle attached to the casting was relatively long, the sinter cake was large, which resulted in the low casting yield of 30% to 40%. Although a longer handle could

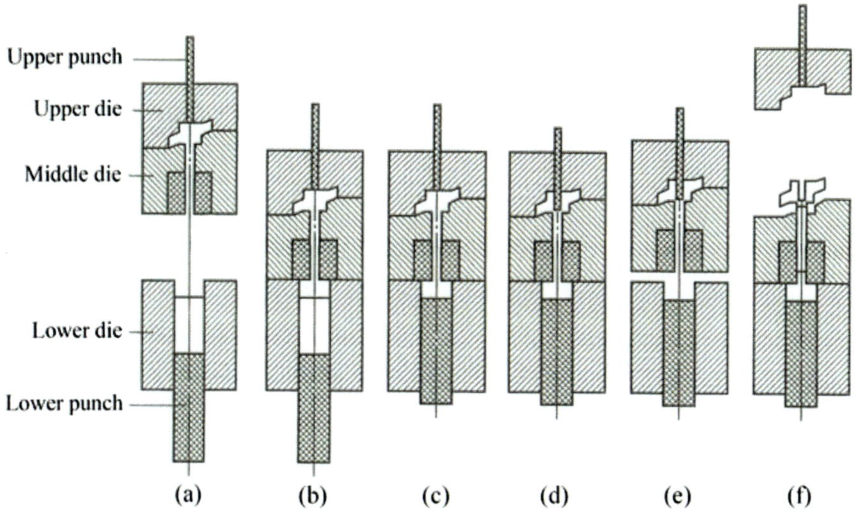

Fig. 6.6 Indirect squeeze casting process: **a** Pouring, **b** Clamping, **c** Filling under the squeezing of lower punch, **d** Squeezing and feeding by the upper punch, **e** Pulling off the handle, **f** Demolding

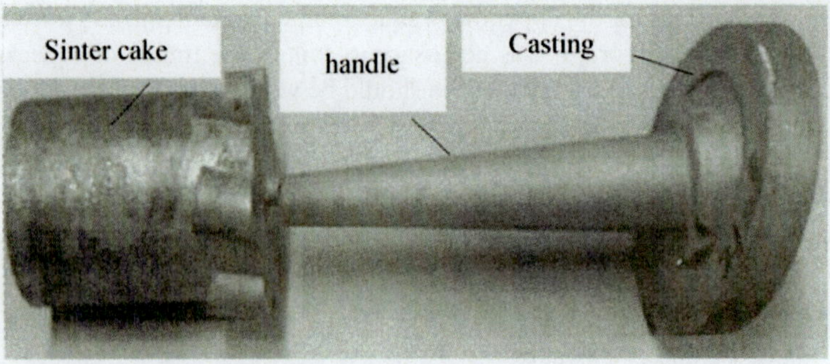

Fig. 6.7 The part obtained by indirect squeeze casting

achieve a good slag retention effect, the structure of the die was more complex, and the long and narrow sprue was not conducive to pressure transmission, resulting in a large pressure loss.

4. Double-Punch Indirect Squeeze Casting with Storage Chamber

The process of the double-punche indirect squeeze casting with storage chamber is shown in Fig. 6.8. Compared with the indirect squeeze casting, it saved the long and narrow sprue. The outside of the material chamber was provided with a thermal insulation sleeve, so that when the aluminum alloy melt was poured into the inside, a certain thermal insulation effect could be obtained, and the thickness of the shell layer would be reduced. The aluminum alloy melt was driven by the lower punch, the filling process was very stable, and the possibility of entrainment was greatly reduced. Meanwhile, the melt with uniform temperature could fill the cavity in a short time, avoiding the uneven chilling effect of the low-temperature die wall on it,

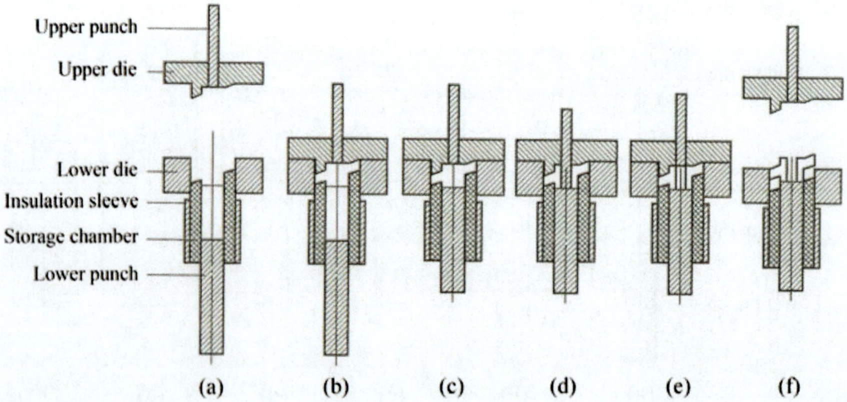

Fig. 6.8 Double-punch indirect squeeze casting process with storage chamber: **a** Pouring, **b** Clamping, **c** Filling under the squeezing of lower punch, **d** Squeezing and feeding by upper punch, **e** Extraction of upper punch, **f** Demolding

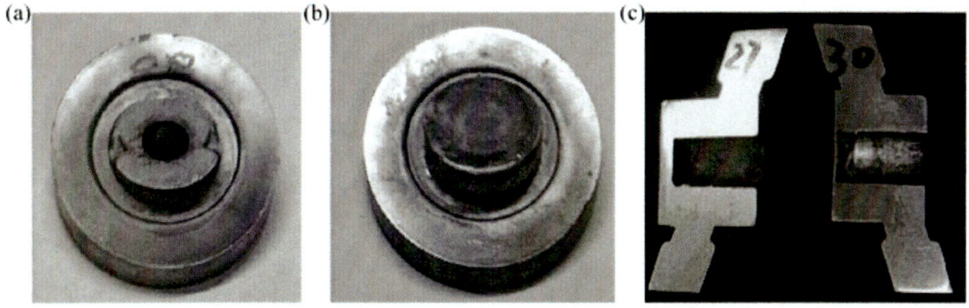

Fig. 6.9 The casting of swash plate: **a** Front side, **b** Back side, **c** Section

and improving the uniformity of cooling. There was no cold partition on the surface of the formed part and no shrinkage cavities or porosities inside, as shown in Fig. 6.9.

Because the die did not have a slag stop device, the casting was prone to form the defect of cold interlayer. The temperature of the inner wall of the storage chamber was usually only 200 ~ 300°C. When the aluminum alloy melt of about 800°C was poured into the chamber, a 0.2 ~ 0.6 mm chilled layer was formed on the surface of the chamber because of the rapid heat conduction of the chamber. As the aluminum alloy melt was squeezed into the die cavity by the lower punch, part or all of the chilled layer enterred the die cavity together with the aluminum alloy melt and formed a cold interlayer in the casting. This kind of cold interlayer was different from cold partitions and inclusions in general castings. It was a layer of chilled structure with extremely fine grains. The cold interlayer caused by the chilled layer existed in the casting in an independent form which split the matrix like a crack, and destroyed the continuity of the matrix. There were four methods for eliminating cold interlayer: (1) Keeping the aluminum alloy melt with a certain degree of superheat to remelt the chilled layer when it was squeezed into the die cavity, (2) Properly increasing the die temperature to reduce the thickness of the chilled layer, which is beneficial to the remelting and breaking of the chilled layer; (3) Increasing the pouring temperature of aluminum alloy and prolonging its liquid time to eliminate the formation of cold interlayer; (4) Properly increasing the filling velocity of aluminum alloy melt to eliminate or reduce the formation of cold interlayer.

Compared with the indirect squeeze casting, the quality of the formed castings was better, and the casting yield was higher (>90%). Although the aluminum alloy melt was easily brought into the coating and other slag inclusions during the filling process, which has a certain impact on the mechanical properties of the casting, the double-punch squeeze casting with storage chamber has more advantages compared to the other three schemes.

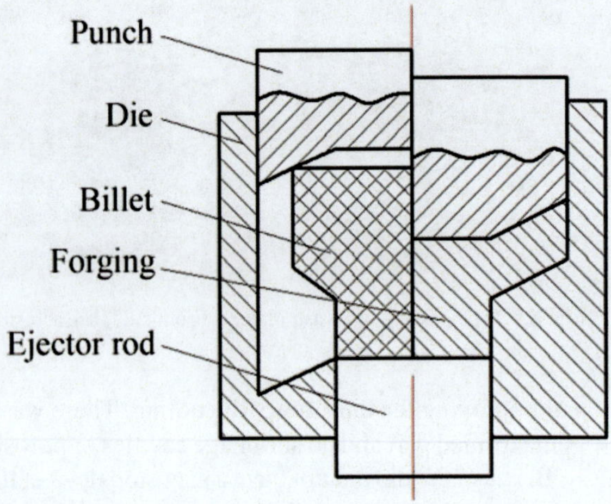

Fig. 6.10 Schematic of closed precision forging of swash plate

6.2.2 Closed Precision Forging Process

The principle of closed precision forging is shown in Fig. 6.10. The billet used for closed precision forging was a stepped bar, which was formed by extruding a Φ65 × 53.6 bar to a Φ38 × 15 rod. Correspondingly, the size of the head became Φ65 × 48.5, and the rod could be used for positioning in the die, as shown in Fig. 6.10. It can be seen that at the beginning of closed precision forging, the deformation was mainly upsetting and partial backward extrusion of bosses. When the backward extrusion was completed, the deformation was completely transformed into upsetting. When the drum was in contact with the side wall of the die cavity, the upsetting was transformed into the radial extrusion with the movement of the punch until the end of forming.

The billet was designed as a bar with a 68.5° cone, which was mainly to reduce the height-to-diameter ratio of the upset part. This part would contact the side wall of the die cavity as soon as possible during upsetting, so the three-dimensional compressive stress was generated inside, which improved the plastic forming performance of Al-Si alloy and avoided the occurrence of cracks.

6.2.3 Comparison of Squeeze Casting and Closed Precision Forging

The A390 aluminum alloy was used as the material, the forming test of the swash plate was carried out by the double-punch squeeze casting process with storage chamber and the closed precision forging process, respectively. When the double-punch squeeze casting with storage chamber was employed, 1.0% (wt) P-10Cu was added at 830 °C for modification. The specific pressure of squeeze casting was

120 MPa, the pouring temperature was 820 ~ 835°C, and the die preheating temperature was 150 ± 10°C. T6 heat treatment was carried out for the casting, and its process parameters were 490°C × 8 h for solid solution and 180°C × 8 h for aging. In closed precision forging, the billet was heated to 480 °C, the die was preheated to 200 °C. The hydraulic press was used as the forging equipment.

Figure 6.11 shows the metallographic micortructure of the forging and squeezed casting. It can be seen that the size of the primary silicon of the forging and squeezed casting was below 30 μm, and the eutectic silicon was finely dispersed and evenly distributed. The mechanical properties of the two types of parts were listed in Table 6.1, which indicates that the tensile strength of squeezed casting was close to that of forging.

Air-conditioning compressors are a series of products with multiple specifications. Correspondingly, the swash plate is also a series of products with different sizes. It can be inferred that the squeeze casting process is more suitable for the production of swash plates with small size. This is because the pressure transmission attenuation of squeeze casting is small and it is easy to obtain a microstructure without defects. Closed precision forging is suitable for the production of swash plates of all specifications, and its advantages are more significant when producing swash plates with large size. This is because closed precision forging can improve the plastic forming performance of the material through a large degree of deformation and obtain forgings with refined and uniform primary silicon.

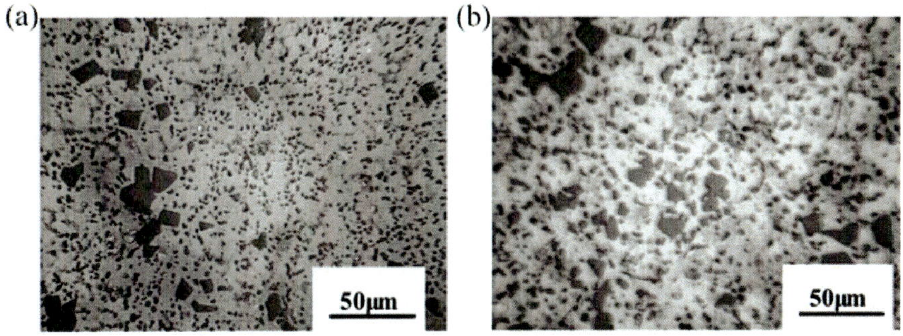

Fig. 6.11 Microstructure of the forging **a** and squeezed casting **b**

Table 6.1 Mechanical properties of two types of parts

Formed parts	Brinell hardness (HB)	Tensile strength (MPa)
Forging	140	342
Squeezed casting	146	334 s

6.3 Precision Forging Process of the Compressor Piston

6.3.1 Precision Forging Process

The dimensions of the piston forging and three-dimensional modeling are shown in Figs. 6.12 and 6.13, respectively. The shape of the forging was symmetrical about the center planes. The cross-sectional area along the axis direction varied greatly near the two ends. It was large at both ends and small in the middle. The minimum cross-sectional area was at the length of 3.6 mm from the middle, which was approximately 904.2 mm^2. The maximum cross-sectional area at the two ends of the forging was 178.3 mm^2. The difference between the two values was more than 4 times. The length with the largest cross section accounted for 1/5 of the total length. The volume distribution in the middle of the forging about 4/5 of the length was relatively uniform.

The uneven distribution of the cross-section along the axis direction caused the metal to deform drastically and unevenly during the forging process, which not only increased the forging force and aggravated the wear of the die, but may also caused defects such as underfilling, folding, and cracking of the forging. According to the cross-sectional distribution characteristics of the piston forging, the designed forging process included three steps: closed pre-forging with wedge-shaped punch → small

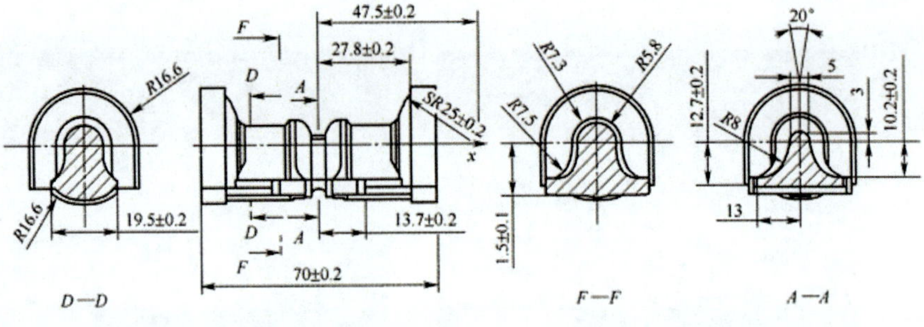

Fig. 6.12 The drawing of piston forging

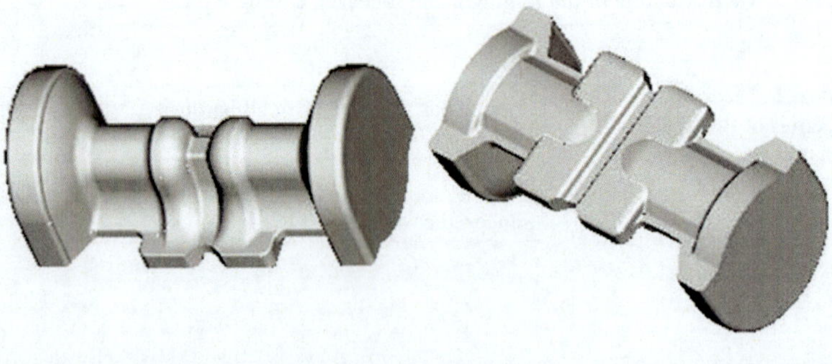

Fig. 6.13 Three-dimensional modeling of the piston forging

flash final forging → trimming. In this study, the cast aluminum alloy bar was cut and used as raw material. In the pre-forging step, the arc profile of the forging was formed to reasonably optimize the distribution of the metal along the axis direction.

6.3.2 *Finite Element Simulation of the Closed Pre-Forging Process*

The pre-forging of the piston was to try to make the material in the middle of the bar flow to both ends under the action of the wedge-shaped working surface of the die to form a reasonable volume distribution along the axis direction. The schematic of the wedge-shaped pre-forging punch is shown in Fig. 6.14.

The pre-forging was completed in a closed die cavity composed of the punch and the die. The working surface of the punch was a wedge-shaped surface with a certain slope. Under the action of this surface, the metal in the middle of the bar was easier to flow to both sides, so that more metal was gathered at both ends of the billet to facilitate final forging. In addition, closed forging was conducive to the formation of strong three-dimensional compressive stress in the die cavity and improved the plasticity and fluidity of the material. During closed precision forging, the material finally reached the flow control chamber, which not only reduced the forging force, but also avoided the formation of large longitudinal burrs.

Figure 6.15 shows the finite element simulation model of the pre-forging process. Since the forging was symmetrical about the center planes, 1/4 of the parts was taken for the simulation to reduce the amount of calculation. The billet was divided into 10,000 units, and the punch and die were both divided into 15,000 units. Figure 6.16 shows the strain distribution of the pre-forging process. It can be seen that under the action of the wedge-shaped punch, the metal in the middle of the billet could flow to both ends relatively easily, and the plastic deformation zone of the billet was larger, so that the metal of the billet was distributed along the axis more reasonably. The deformation zone of the billet could be divided into three zones: I, II and III. The

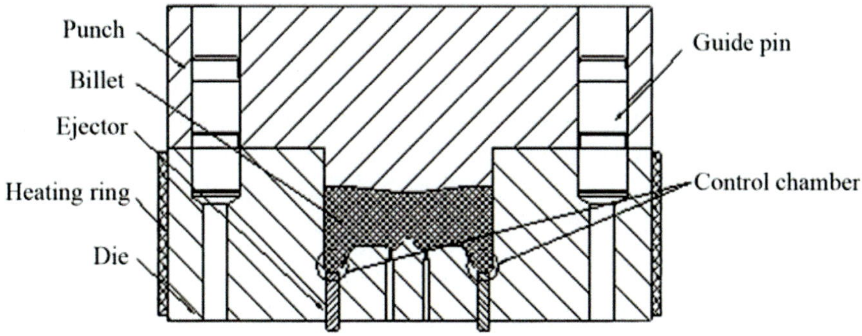

Fig. 6.14 Schematic of the pre-forging die

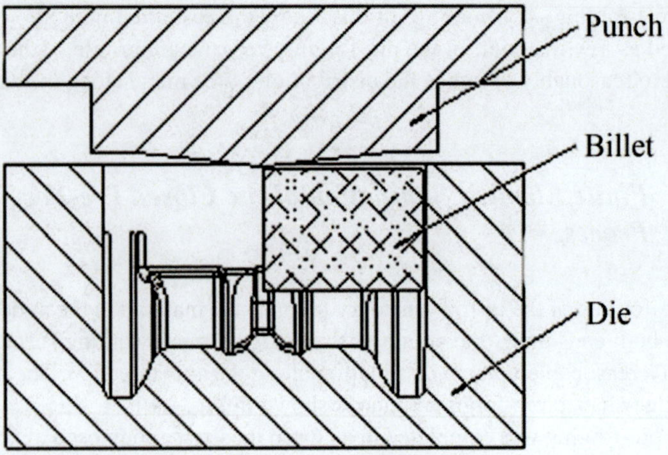

Fig. 6.15 Finite element simulation model of the piston

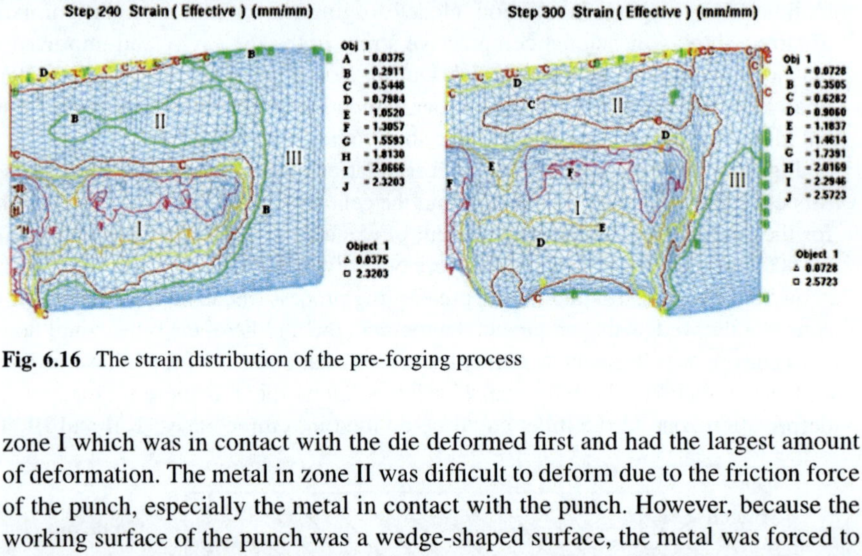

Fig. 6.16 The strain distribution of the pre-forging process

zone I which was in contact with the die deformed first and had the largest amount of deformation. The metal in zone II was difficult to deform due to the friction force of the punch, especially the metal in contact with the punch. However, because the working surface of the punch was a wedge-shaped surface, the metal was forced to flow to both ends. The metal in the middle of the billet flowed to both ends, and a large amount of metal accumulated in the cavities at both ends, causing a large upsetting deformation in zone III.

6.3.3 Process Test

According to the above process analysis and finite element simulation, the pre-forging and final forging of piston were carried out. The pre-forging die and the final forging die were arranged on a set of die holders, as shown in Fig. 6.17.

The pre-forging die and final forging die were designed with a combined structure, which facilitated the assembly, debugging and maintenance of the die. In order to avoid the damage of forging profile surface during ejection, the ejector rods were arranged at both ends of the die cavity. Figure 6.18 shows the pre-forged part and final forging part. And Fig. 6.19 shows the forging after trimming. The surface accuracy and dimensional accuracy of the forging part fully met the design requirements.

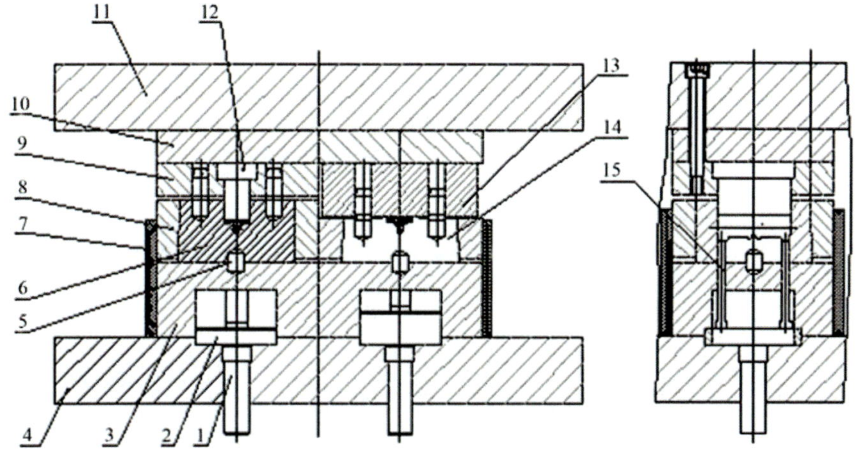

Fig. 6.17 The forging die of the piston. 1-Ejector rod; 2-Backing plate for ejector; 3-Backing plate for die; 4-Lower template; 5-Pin 6-Pre-forging die; 7-Heating ring; 8-Fixed ring; 9-Fixed plate; 10-Backing plate for punch; 11-Upper template; 12-Pre-forging punch; 13-Final forging punch; 14-Final forging die; 15-Ejector

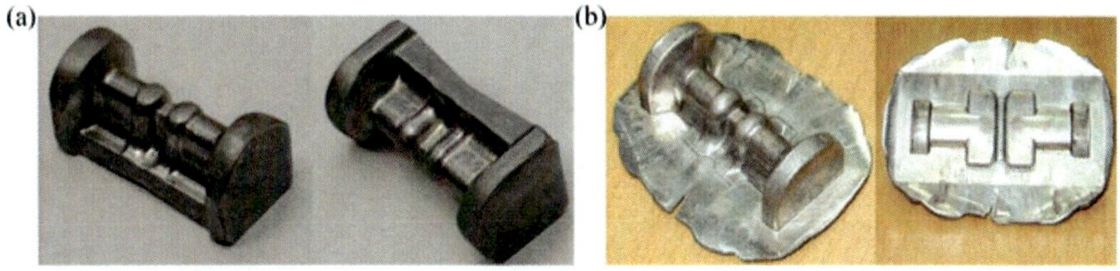

Fig. 6.18 The pre-forged part **a** and final forged part **b**

Fig. 6.19 The forgings after trimming

6.4 Combined Casting-Forging of A356 Aluminum Alloy Steering Knuckle

Figures 6.20 and 6.21 show the three-dimensional modeling and a photo of the steering knuckle of automobiles. The outline size was 291 mm × 264.3 mm × 120 mm with a Φ90mm round hole in the middle. The part had multiple bosses and small round holes for connecting with other parts. Based on the structural analysis of the steering knuckle, and the material was also A356 aluminum alloy, the combined casting-forging process was proposed for the precision forming of the part. The forming process of the steering knuckle was as follows: aluminum alloy ingot melting → squeeze casting → removal of handles and feeders → heating → small flash precision forging → trimming → heat treatment → machining. The squeeze casting process of the steering knuckle was similar to that of swash plate [7, 8], so the description will not be repeated here.

Figure 6.22 shows a schematic of the small flash precision forging process for the steering knuckle. The dotted line in the figure is the squeeze casting, and the solid line is the forging. During forging process, the casting underwent upsetting deformation. The drawing of cross-section A-A shows the relationship between the cross-section of the casting in any vertical direction on the steering knuckle and the

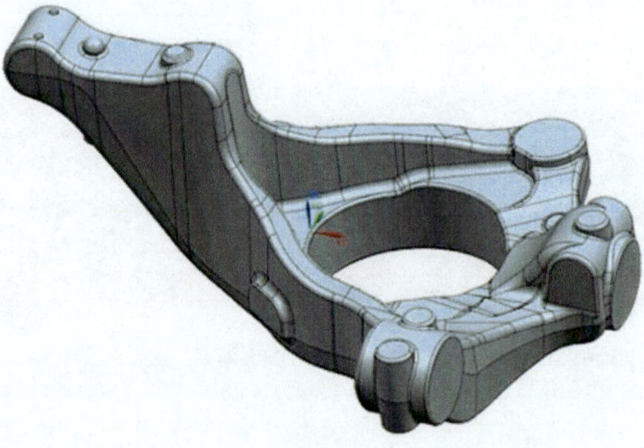

Fig. 6.20 Three-dimensional modeling of the steering knuckle

Fig. 6.21 The photo of steering knuckle

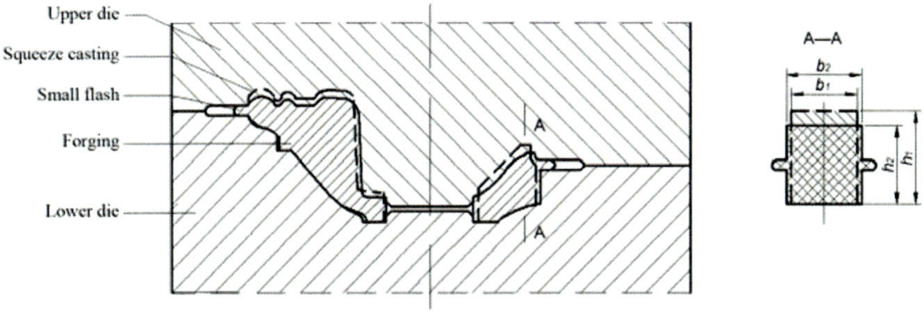

Fig. 6.22 Schematic of small flash precision forging process for steering knuckle

cross-section of the forging and the small flash.

$$A_1 = h_1 \cdot b_1 = A_2 + 2A_\mathrm{f} = h_2 \cdot b_2 + 2A_f$$

where A_1, h_1 and b_1 are the cross-sectional area, height and width of the casting, respectively; A_2, h_2 and b_2 are the cross-sectional area, height and width of the forging, respectively.

The relationship between the section height and width of the casting and the forging should be:

$$h_1 > h_2, b_1 < b_2$$

The ratio of h_1/h_2 or b_2/b_1 was the key to realize the squeeze casting and small flash precision forging process, that is, the upsetting ratio was positively related to the density and mechanical properties of the forgings. The purpose of precision forging with small flash was to generate strong three-dimensional compressive stress through the thin flash formed around the forging, and improved the plastic forming performance to increase the density of the forging, thereby improving the mechanical properties of the forging.

During the combined casting-forging process, the metal had good fluidity in the molten state, and could be formed into complex steering knuckle. The small flash precision forging could effectively eliminate internal defects of the casting, and improved mechanical properties. After the handles and feeders of the casting were cut off, the temperature of the casting might be maintained at about 450 °C, so the casting can be directly formed by small flash hot precision forging. In this way, the heating process was reduced, which not only saved heating energy consumption, but also improved production efficiency. The handles and feeders cut from the castings and the flash cut from the forgings could be melted and reused, so the material utilization was significantly improved.

6.5 Combined Casting-Forging of A356 Aluminum Alloy Wheel [9]

6.5.1 Process Plan

There is another combined casting-forging forming process, which uses the same set of dies to realize the process of casting first and then forging. The process is also known as double squeeze casting. The process includes four steps, as shown in Fig. 6.23. Firstly, the injection piston squeezes the alloy liquid into the die cavity; then the alloy liquid crystallizes and solidifies under the pressure of the injection piston; subsequently, the cast part is forged by the punch (the degree of deformation is usually small); finally, the forged part is ejected from the die.

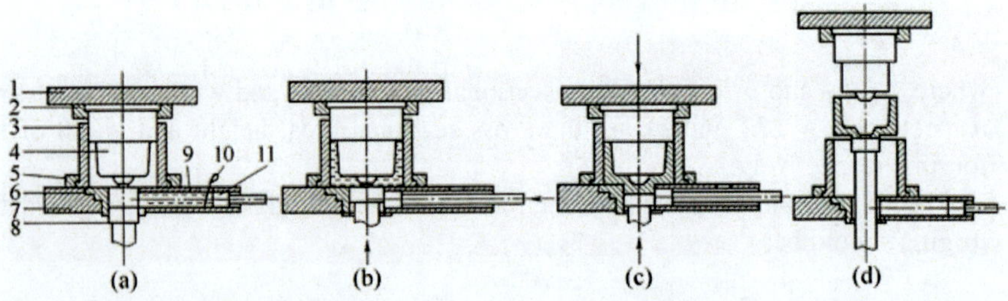

Fig. 6.23 Combined casting-forging process: **a** Pouring, **b** Solidification, **c** Forging, **d** Ejection. 1-Upper template; 2-Upper pressing plate; 3-Die 4-Punch; 5-Lower pressing plate; 6-barrel; 7-Lower template; 8-Injection piston; 9-Transporting pipe for molten liquid; 10-Quantitative spoon; 11-Pressure head of transporting pipe

6.5.2 *The Effect of Process Parameters on the Wheel Forming*

In the combined casting-forging process, the die temperature, piston injection pressure, and forging time are important influencing factors. The numerical simulation of the forming process was employed to analyze the influence law of each parameter. The forging zones were the wheel core and spokes. Assuming that the alloy liquid was cooled in the die. The pouring temperature was 700 °C, and the initial forging temperature was 480 °C. The forging reduction was 3 mm.

The typical distribution of the temperature field of the wheel when the die temperatures was 200 °C is shown in Fig. 6.24. It can be seen that the cooling rate of the thick-wall was slower, and defects such as shrinkage porosity and shrinkage holes were prone to appear in this place. With the increase of the die temperature, the forging force decreased significantly. Therefore, a higher die temperature within a certain temperature range could reduce the forging force and save energy.

After the melt filling, it was necessary to apply a certain pressure to the melt by the injection piston to obtain a high-density wheel. Since the pressure directly affects the densigy and uniformity of the formed microstructure, the stress under different pressures was simulated and analyzed. The stress in the wheel core and spokes was higher than that of the other zones, which was beneficial for obtaining denser microstructure in these zones. As the piston pressure increases, the three-dimensional compressive stress on the workpiece increased significantly. However, when the piston pressure exceedde 150 MPa, the three-dimensional compressive stress on the part increases slightly, but the material was extruded from the contact surface of the die to generate flash, which could lead to die deformation or damage. Therefore, the piston ressure should be 100 MPa ~ 150 MPa.

At different forging velocities, the required forging force was also different. The forging force under different forging velocities is shown in Fig. 6.25. Under the same process conditions, the smaller the forging velocity, the smaller the forging force. Due to the high die temperature, the temperature had little effect on the forging force, so the forging velocity was the main factor that affected the forging force. It should be

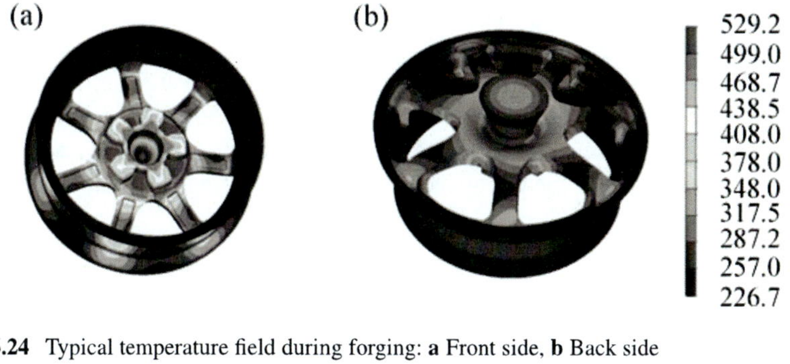

Fig. 6.24 Typical temperature field during forging: **a** Front side, **b** Back side

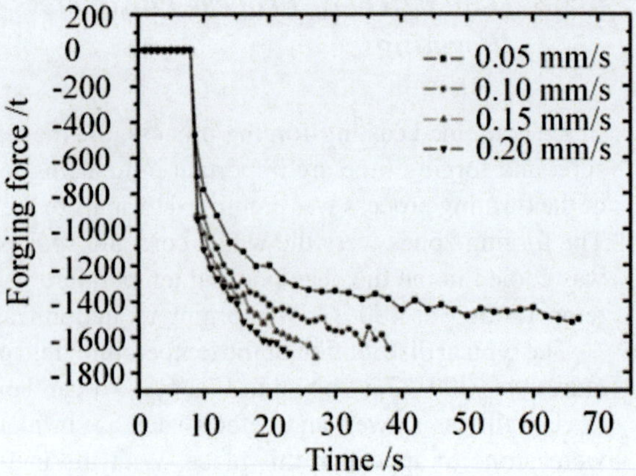

Fig. 6.25 The forging force under different forging velocities

pointed out that too small forging velocity was not conducive to the forming of wheel, and a larger forging velocity may cause the metal to locally overheat. Therefore, the forging velocity of the wheel was 0.1/mm/s ~ 0.15 mm/s.

6.5.3 Process Test

A forming experiment of the A356 aluminum alloy wheel was carried out, and the microstructure was analyzed. An SCV ~ 2000 vertical squeeze casting machine was used for casting. The experimental parameters were set according to the simulation results, namely, the pouring temperature was 700°C, the die temperature was 300 °C, the forging reduction was 3 mm, the forging velocity was 0.12 mm/s, the piston pressure was 100 MPa, and the forging started after 8 s of the filling. The formed aluminum alloy wheel is shown in Fig. 6.26.

The metallographic samples were taken from the connection zone between the spokes and the rim in the non-forged zone of the hub and the hub boss in the forged zone. The metallographic structure is shown in Fig. 6.27.

The microstructure at the connection zone of the spokes and the rim was dominated by a dendritic structure of α - Al which was a typical casting structure. The shrinkage porosity of this zone was 0.5%. The casting structure at the hub boss had been transformed into forging structure, the shrinkage porosity became 0.3%. This was consistent with the above results, indicating that the combined casting-forging process could improve the micortructure and the performance of the wheel.

Fig. 6.26 The formed wheel and sampling position

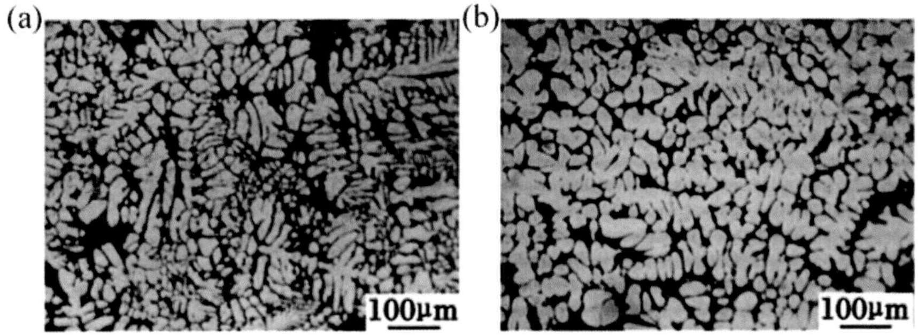

Fig. 6.27 Microstructure of the wheel samples: **a** The unforged zone, **b** The forged zone

References

1. Muyali S, Yong MS (2010) Liquid forging of thin Al-Si sturctures. J Mater Process Technol 210(10):1276–1281
2. Cheng JW, Xia JC, Dai HM, Hu GA (2006) Analysis of die forging process for the piston tail part and numerical value simulaiton. Forging & Stamping Technology 31(4):1–4
3. Dai HM, Cheng JW, Xia JC, Hu GA (2006) Experimental study on forging process of piston tail part. Die & Mould Industry 32(7):60–64
4. Cheng JW (2006) Study on the key technology of precision die forging for high strength Al-alloy parts in car. Huazhong University of Science and Technology, Wuhan
5. Deng L (2011) Applied fundamental study on precision forging of aluminum alloy. Huazhong University of Science and Technology, Wuhan
6. Lan GD (2007) Research on squeeze casting technology of high-silicon aluminum alloy parts. Huazhong University of Science and Technology, Wuhan
7. Tu WJ, Wang G (2015) Study on squeeze casting processing for automotive Al-alloy steering knuckle. Foundry 64(8):740–743

8. Kim M, Lim T, Yoon K, Ko Y, Kim KJM, Kwak K (2011) Development of cast-forged knuckle using high strength aluminum alloy. SAE Technical Paper, 2011, 2011–01–0537
9. Zhang Q, Cao M, Zhao SD, Liu Q (2014) Integrated casting and forging process for aluminum alloy automobile wheel. J Plast Eng 21(2):1–6

Chapter 7
Precision Forging Presses for Aluminum Alloy

7.1 Overview

7.1.1 The Requirements of Precision Forging on Equipment

Precision forging equipment for aluminum alloy is developed on the basis of general die forging equipment according to different aluminum alloy properties and structural characteristics of aluminum alloy parts. At present, for aluminum alloy precision forging, general die forging equipment is more commonly used. In recent years, some new precision forging equipment, as well as automated production technology have also been developed.

Whether it is a general die forging equipment or a new forging equipment, it should meet the requirements of the precision forging process. The main requirements includes high rigidity of press body, high guiding accuracy between the slider and the guide rail of the press body, high movement velocity of slider, quick ejection and enough installing space for the die.

The rigidity of the press body refers to the ability of the press body to resist elastic deformation during the process of precision forging. The higher the rigidity, the smaller the elastic deformation, the higher the dimensional accuracy of the forgings in the height direction, and the better the height and dimensional consistency of the mass-produced forgings. The higher the guiding accuracy, the higher the coaxiality of the punch and the die during precision forging. Therefore, the dimensional accuracy of the forgings in the horizontal direction and the consistency of the horizontal dimensions of the forgings in mass production will be better.

For mechanical presses and screw presses, the slider velocity refers to the average velocity of the slider. For hydraulic presses, the slider velocity includes the idle velocity of the slider and the forging velocity. Generally, the forging temperature range of aluminum alloy is not more than 100 °C. On the one hand, a higher slider velocity is beneficial to avoid excessive cooling of the billet and to ensure that the

L. Deng et al., *Precision Forging Technology and Equipment for Aluminum Alloy*, Springer Series in Advanced Manufacturing,
https://doi.org/10.1007/978-981-19-1828-5_7

metal is formed at a higher temperature during hot precision forging. On the other hand, it is beneficial to improve production efficiency.

During hot precision forging, the heat of th high-temperature billet was transferred to the die, which increases the die temperature. The longer the forging stays in the die, the higher the die temperature will be, which will easily lead to wear of the die. Therefore, after die forging, the forging should be ejected from the die quickly. In addition to the fast ejection velocity, the ejection device should be linked with the slider of the press. When the slider of the press returns, the ejection device immediately eject the forging from the die.

For closed precision forging dies, especially separable dies, because of the addition of a die clamping mechanism, the vertical and horizontal dimensions are significantly enlarged. Therefore, when selecting or designing precision forging equipment, it should be considered that whether the size of the installing space for the die is sufficient.

7.1.2 Types of Precision Forging Equipment

At present, there are two major types of die forging equipment for aluminum alloy: general die forging equipment and special die forging equipment. General die forging equipment includes mechanical presses, screw presses, single-action hydraulic presses, etc. Special die forging equipment includes various types of isothermal die forging hydraulic presses, multi-action hydraulic presses, multi-directional hydraulic presses and large hydraulic presses. According to the transmission method, these two types of equipment can be divided into many different categories. Table 7.1 lists the classifications of various die forging equipment that can be used for aluminum alloy precision forging.

7.1.3 Selection of Precision Forging Equipment for Aluminum Alloy

By equipping dies with with precise guiding devices, the general die forging equipment can meet the requirements of aluminum alloy precision forging. In this way, the cycle of preparing for precision forging production can be shortened, and the cost of equipment investment can be reduced. In addition, the forging process of aluminum alloy requires not only a fast forging velocity, but also a high idle velocity. Based on the general hydraulic press and electric screw press, the equipment can be modified to meet the requirements of precision forgings.

In view of the special characteristics of cast aluminum alloy and high-strength aluminum alloy, it is difficult for the existing general equipment to meet the precision forging requirements even after modification. Therefore, new precision forging

Table 7.1 Classifications of various equipment used for aluminum alloy precision forging

Type	Category	Subcategory	Transmission method
General die forging equipment	Mechanical press	Universal crank press	Crank connecting rod drive
		Toggle press	Crank connecting rod drive
		Hot die forging press	Wedge transmission, Double slider transmission
	Screw press	Clutch type screw press	Belt drive
		Non-direct drive electric screw press	Gear drive
		Direct drive electric screw press	Electrode rotor drive
	Single-action hydraulic press	Die forging hydraulic press	Hydraulic transmission
Special die forging equipment	Hydraulic press	Isothermal forging hydraulic press	Hydraulic transmission
		Multi-action hydraulic press	Hydraulic transmission
		Multi-directional hydraulic press	Hydraulic transmission
		Large hydraulic press	Hydraulic transmission
		Squeeze casting hydraulic press	Hydraulic transmission

equipment should be developed, for example, new high speed precision forging hydraulic presses and direct drive servo presses.

7.1.4 Force and Energy Characteristics of General Die Forging Equipment

When selecting die forging equipment, the forging force P_P and forging energy E_P are the most important factors to be considered. The forging force generally refers to the maximum deformation force required to form qualified forgings. The forging energy refers to the work done by the forging force in the working stroke, which can be determined by the area enclosed by the force–stroke curve. Figure 7.1 shows the force–stroke curves of different forging processes. Among them, m_f is the characteristic coefficient, which represents the ratio of the shaded area to the area enclosed by the force–stroke curve.

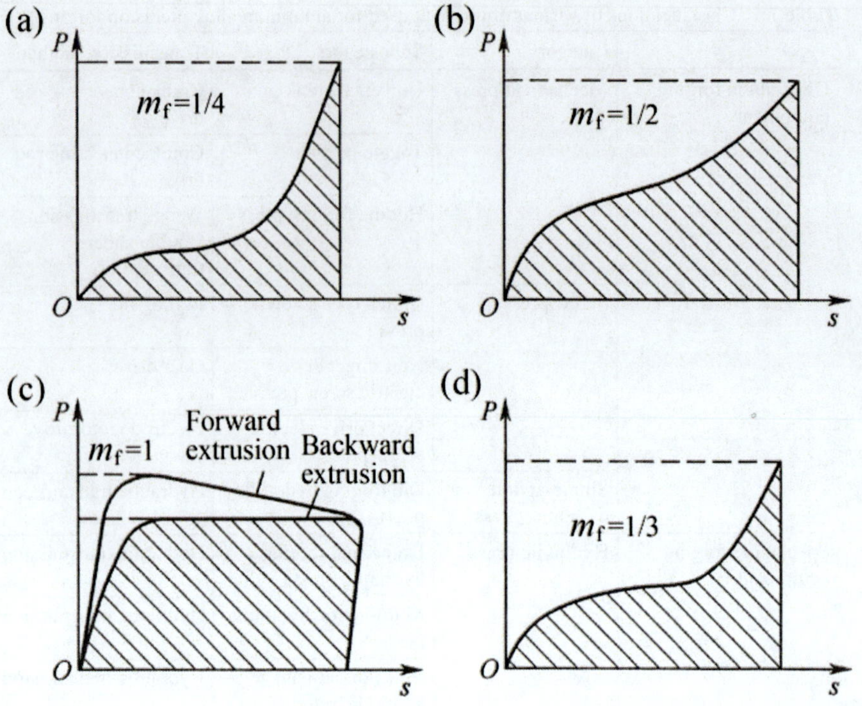

Fig. 7.1 Force–stroke curves of different forging processes: **a** Open die forging, **b** Free upsetting, **c** Forward extrusion and backward extrusion, **d** Closed precision forging

The force and energy that can be generated by die forging equipment when it is working is called effective force P_M and effective energy E_M. Reasonable selection of die forging equipment should ensure the following conditions, namely

$$P_M \geq P_P \tag{7.1}$$

$$E_M \geq E_p \tag{7.2}$$

That is to say, the effective force of the equipment should be greater than or equal to the maximum forging force at any time in the precision forging process, and the effective energy of the equipment should be greater than or equal to the forging energy. If Eq. (7.1) is not satisfied, the mechanical press will be overloaded, causing damage to the press body and dies, and the hydraulic press will stop without reaching the predetermined deformation. If Eq. (7.2) is not satisfied, the screw press will not be able to obtain qualified forgings in one stroke.

It can be seen that when selecting die forging equipment, it is necessary to understand the force and energy characteristics of various die forging equipment. The following is a more detailed introduction to the general equipment and some new precision forging equipment.

7.2 Hot Die Forging Press

7.2.1 Characteristics of Hot Die Forging Press

Multi-station forging processes can be performed on the hot die forging press. The forging has high dimensional accuracy and small machining allowance, but the phenomenon of jam may occur during forging. Therefore, in addition to the requirements described in Sect. 7.1, the hot die forging press has the other three characteristics in terms of structure and performance, including strong anti-tilting ability of slider, more motion beats of slider and device to relieve jam.

If the slider cannot resist tilt under the action of a large eccentric load, the thickness of the forging will be uneven. To improve the anti-tilting ability of the slider, methods such as reducing the gap of the guide rail, increasing the guide length of the slider, strengthening the axial rigidity of the guide rail, and increasing the width of the crank can be employed.

The more motion beats can reduce the residence time of the hot die forgings in the forging die. This is not only conducive to the forming of the part, but also conducive to extending the life of the die.

Since the hot die forging press uses a rigid transmission and has a fixed bottom center, when the billet size is too large, or the billet temperature is too low, the phenomenon of jam may occur. Therefore, the press should have a device for unloading its working mechanism.

7.2.2 Basic Structure and Working Principle of Hot Die Forging Press

1. Hot die forging press of connecting rod type

The structural diagram of the connecting rod type press is shown in Fig. 7.2 [1]. It can be seen that the press employs a crank-slider working mechanism, and has two-stage transmission including belt transmission and gear transmission. The clutch and brake are installed on the left and right ends of the low-velocity eccentric shaft respectively and pneumatically interlocked. The slider employs the type of elephant trunk, which has an additional guiding surface to improve the anti-tilting ability of the slider. The adjustment of the shut height is by means of a double-wedge workbench.

The structural features of this type of press are as follows:

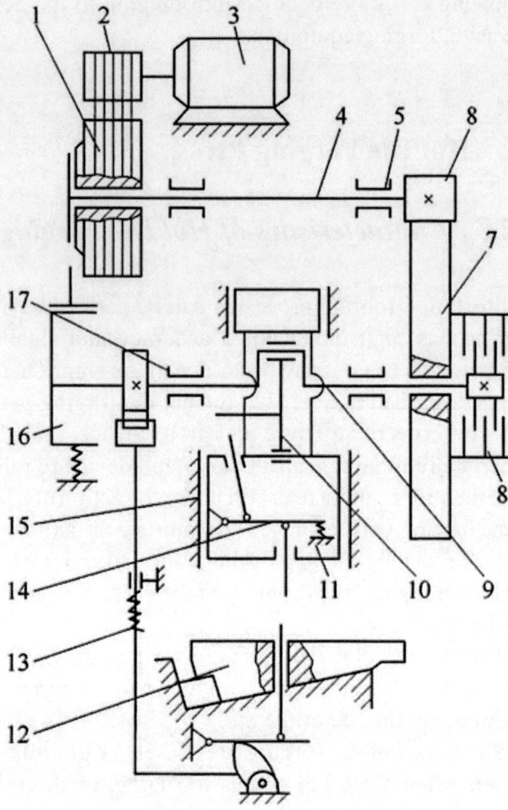

Fig. 7.2 Structural diagram of the connecting rod type press. 1-Big belt pulley; 2-Small belt pulley; 3-Motor; 4-Drive shaft; 5, 17-Bearing; 6-Small gear; 7-Big gear; 8-Clutch; 9-Eccentric shaft; 10-Connecting rod; 11-Slider; 12-Wedge-shaped workbench; 13-Lower ejection device; 14-Upper ejection device; 15-Guide rail; 16-Brake; 17-Sliding bearing

(1) The integral solid casting is employed to manufacture the frame. Compared with the hollow or combined frame, the integral solid frame has better longitudinal and lateral rigidity, so the precision of the forging is higher. For the large frames that exceed the transport limit, the combined frame with pre-tightened tie rod is also used. There are windows on both sides of the frame, the width and height of which can be desiged according to the automatic transfer system for the part.

(2) The eccentric shaft is made of alloy steel forgings. A reasonable arc is designed at the transition fillet of the eccentric shaft to reduce stress concentration.

(3) The connecting rod is made of alloy steel castings. The width of the connecting rod and the eccentric shaft is equal to the width of the slider, which can bear larger eccentric load and improve the anti-tilting ability of the press.

(4) The slider employs a rectangular structure, and the guide rail employs a diagonal type. If the requirements for the forging accuracy are high, a water-cooling mechanism can also be used to maintain the precise clearance of the guide rail.

(5) Highly reliable and durable friction clutches and friction brakes are installed on both ends of the eccentric shaft. The friction brake is equipped with a

water cooling system, which can ensure stable operation under harsh working conditions. The friction element employs a floating friction block which is convenient for maintenance and replacement.

2. Hot die forging press of wedge type

Figure 7.3 shows a structural diagram of a wedge type press. Compared with the connecting rod structure, this structure has the advantages of smaller total elastic deformation, larger anti-titling ability and lateral rigidity, larger eccentric load bearing capacity, and smaller load on the guide rail. In addition, it is convenient to adjust the height of the die assembly. Because of these advantages, the press is mainly used for precision forging of slender forgings, eccentric rods, and multi-station die forgings with strict requirement of tolerances.

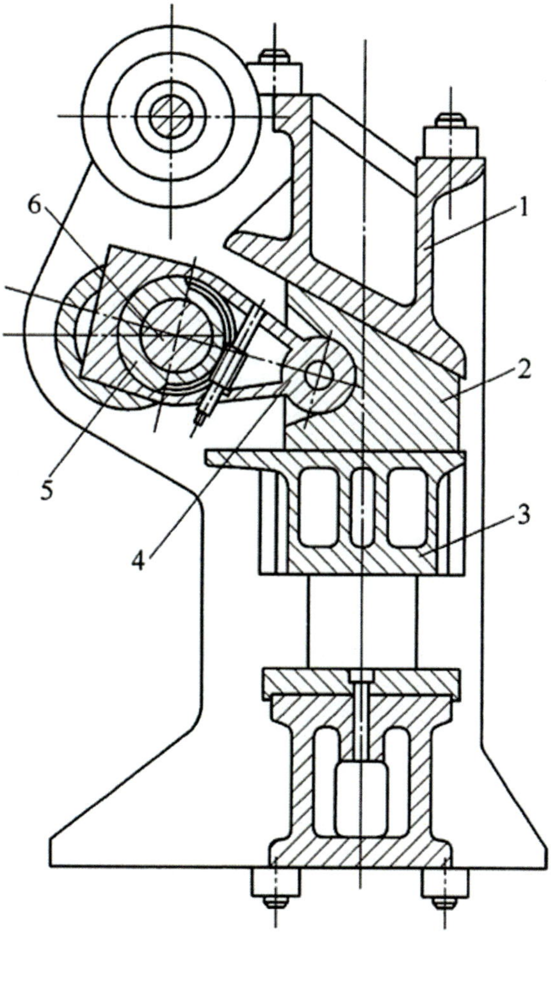

Fig. 7.3 Structural diagram of wedge type press. 1-Press body; 2-Transmission wedge; 3-Slider; 4-Connecting rod; 5-Eccentric worm gear; 6-Crankshaft

The transmission system of the press is that the motor drives the intermediate shaft through a belt. At one end of the intermediate shaft, there is a flywheel. And at the other end is a small gear that meshes with a big gear mounted on a crankshaft. A friction clutch is installed in the big gear, and a brake is installed on the other side of the crankshaft. The clutch and the brake are connected with compressed air. Due to the sturdy structure of the press, higher switching frequency between clutch and brake can be obtained.

The crankshaft transmission mechanism is installed on the rear side of the press. The connecting rod between the crankshaft and the transmission wedge is designed according to a pressing rod. In the crankshaft bearing of the pressure rod, an eccentric sleeve is installed to adjust the stroke position of the slider. The transmission wedge is connected with the frame and the slider to form a closed structure. The balance cylinder keeps the contact surface of the slider and the wedge in a gap-free fit.

7.2.3 Calculation of the Nominal Force of Hot Die Forging Press

(1) For circular forgings,

$$F = 8(1 - 0.001D)\left(1.1 + \frac{2D}{D}\right)^2 A\sigma_s \tag{7.3}$$

(2) For non-circular forgings,

$$F = 8(1 - 0.001D)\left(1.1 + \frac{2D}{D_1}\right)\left(1 + 0.1\sqrt{\frac{L}{B}}\right)A\sigma_s \tag{7.4}$$

where F is the nominal force of the press; A is the horizontal projection area of the forging; L and B are the length and width of the non-circular forging on the horizontal projection surface; σ_s is the yield strength of the alloy at the final forging temperature.

7.2.4 Toggle Type Press

Toggle type press is suitable for precision pressing which is mainly to improve the dimensional accuracy of forgings and reduce the surface roughness. For aluminum alloy forgings, it also can be used for correcting the deformation.

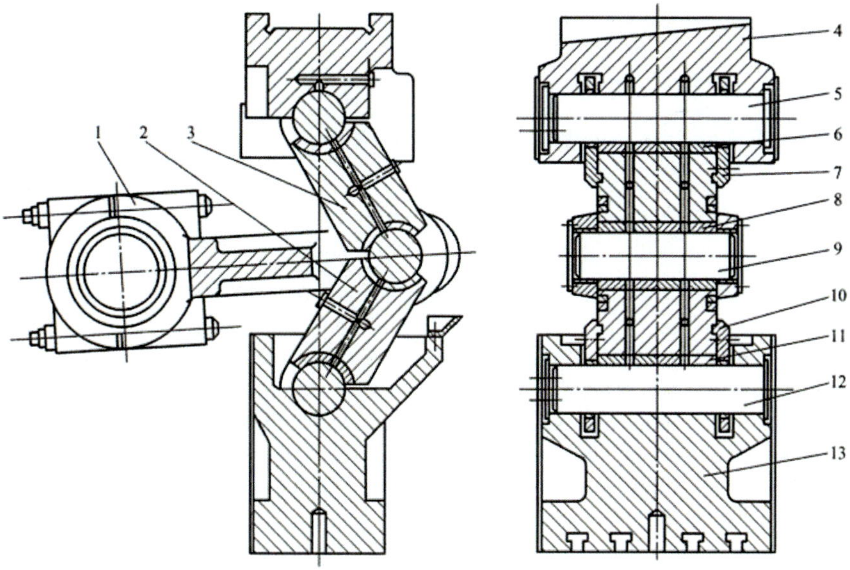

Fig. 7.4 Structual diagram of the toggle mechanism of a typical press. 1-Connecting rod; 2-Lower toggle; 3-Upper toggle; 4-Adjusting slider; 5-Upper toggle; 6-Upper toggle bearing; 7-Upper connecting ring; 8-Middle toggle bearing; 9-Middle toggle shaft; 10-Lower connecting ring; 11-Lower toggle bearing; 12-Lower toggle shaft; 13-Working slider

Figure 7.4 shows the typical structural diagram of the toggle mechanism of a press. It consists of wroking slider, adjusting slider, upper/lower toggle lever, upper/lower connecting ring, upper/middle/lower toggle shaft, upper/middle/lower toggle bearing, and connecting rod. The two sliders are connected by the toggle shaft, toggle bearing, toggle lever and upper and lower connecting rings. The upper and lower toggle levers and the upper and lower connecting rings are respectively connected by bolts. The middle toggle shaft is connected with the crankshaft by a fork-shaped connecting rod. The working slider and the adjusting slider are solid bodies. The upper and lower toggle levers are short and thick structures. Although they bear all the nominal force during work, their deformation is very small, therefore, the press is suitable for fine pressing process. Its working principle is similar to that of a connecting rod type press.

Because the toggle mechanism has to withstand a large force, good lubrication is needed. There are oil passages in the corresponding parts of the toggle lever, toggle bearing and slider. The thin oil gear pump is used for lubrication to ensure the toggle bearing of the toggle system is well lubricated.

The pressure required for precision pressing is mainly related to factors such as the type of aluminum alloy, forging temperature and stress state. The forging force can be calculated by the following equation.

$$F = pA \tag{7.5}$$

where p is the average unit pressure; A is the projected area of the forging.

7.3 Precision Forging Hydraulic Press

7.3.1 *Characteristics of Precision Forging Hydraulic Press*

Since the velocity of the hydraulic press is easy to control, it is more suitable for precision forging. The hydaulic press neither has the requirement of fixed closed space, nor is it restricted by insufficient energy. Its working capacity is only limited by the nominal force, so it can be used for forging, stamping and extrusion. Moreover, because the hydraulic press can reach the maximum pressure at any position of the full stroke of the slider, it is especially suitable for the extrusion-type closed precision forging process that requires relatively large deformation and deformation energy.

The hydraulic press is a load limiting equipment, and its working capacity is generated by pumps and accumulators. The effective pressure of the hydraulic press is calculated by Eq. (7.6):

$$P_{\mathrm{M}} = p_L F_a \tag{7.6}$$

where p_L is the pressure of the working fluid; F_a is the area of the working cylinder or working plunger.

In the hydraulic press with direct pump drive, the maximum liquid pressure is directly determined by the pressure of the pumping station system. The consumed energy varies with the deformation resistance of the forging, and the work efficiency is high. The movement velocity of the slider has nothing to do with the deformation resistance, but only with the flow rate of the pump.

In the hydraulic press driven by a pump-accumulator, the high-pressure liquid is stored in the accumulator when it is not working, and the working pressure is supplied by the pump and the accumulator at the same time, so that a large amount of high-pressure liquid can be provided in a short time. Energy consumption has nothing to do with deformation resistance. Regardless of whether forging deformation is required, a fixed force is provided. Therefore, the greater the deformation stroke, the greater the energy consumed. The movement velocity of the slider is related to the deformation resistance. The greater the deformation resistance, the slower the velocity.

7.3.2 *Requirements of Precision Forging on the Hydraulic Press*

The press should have good resistance to eccentric load and sufficient rigidity, so that precise forgings can be obtained. The guide structure of the slider should be able to ensure the required horizontal dimensional accuracy. And the control system should be able to accurately control the stopping accuracy of the slider in order to ensure the dimensional accuracy in the vertical direction.

The hydraulic press is characterized by a relatively large force exerted on a small area, so the forging force must be distributed to the frame as evenly as possible to reduce eccentric load and stress concentration. The current advanced precision forging hydraulic press employs a long slide guide mechanism with eight parallel planes, which can reduce the stress caused by the eccentric load. And the guide surface is made into sections which can be easily replaced.

The stopping accuracy of the precision forging hydraulic press is about $\pm$ 0.1 mm which is one order of magnitude higher than that of the free forging hydraulic press. The stroke of the slider includes the accumulated error caused by the deflection of the workbench, the compression deformation of the die, the compression deformation of the slider, and the elongation of the frame. The sum of these deformations varies with the applied forging force. Therefore, in order to obtain good stopping accuracy, it is necessary to automatically correct the stroke of slider according to the changes in the forging force. In addition, due to the hysteresis of the electrical and hydraulic systems, the slider may exceed the specified stroke even when the stop position is set. The excess stroke is related to the pressurization rate when the stop command is given and the lag of the command transmission system. By feeding the pressurization rate back to the control system and reducing it to a certain value, less excess stroke can be achieved. In addition, the compressibility of the liquid in the high-pressure pipeline of a hydraulic press has a significant impact on the response time of the hydraulic valve. Therefore, the length of the high-pressure pipeline should be minimized.

The die temperature has a great influence on the surface quality of the forging, the metal flow, and the strength and wear resistance of the die. When the production batch is large, the heat transferred from the billet is sufficient to prevent the die from cooling down, but in the small batch production of forgings with complex shapes and low-plasticity materials, preheating devices and preservation devices are required. Installing a heater in the die base is one of the more common methods. During aluminum alloy forging, the temperature of the die cavity is maintained at 150 °C to 300 °C, so the die base will also rise to a relatively high temperature. In order to reduce the heat transfer to the hydraulic press frame, heat insulation panels and water cooling panels should be installed between the die base and the slider and the workbench.

7.3.3 Isothermal Precision Forging Hydraulic Press

Isothermal precision forging is carried out at a very slow deformation velocity while keeping the forging temperature of the material basically unchanged. Generally, the deformation velocity is about 0.5 ~ 5 mm/s.

Figure 7.5 shows a typical structure of isothermal precision forging hydraulic press. The fixed lower beam is the workbench of the press. The upper beam and the hydraulic cylinder are connected together by four columns, and the slider moves along the columns. The die with heating device is installed on the backing plate. In order to realize the fast return of the slider, a return cylinder is provided. In addition, the press should have a locking device. When the hydraulic system is closed, the slider stops at the top. The locking device is composed of a screw and a nut. When

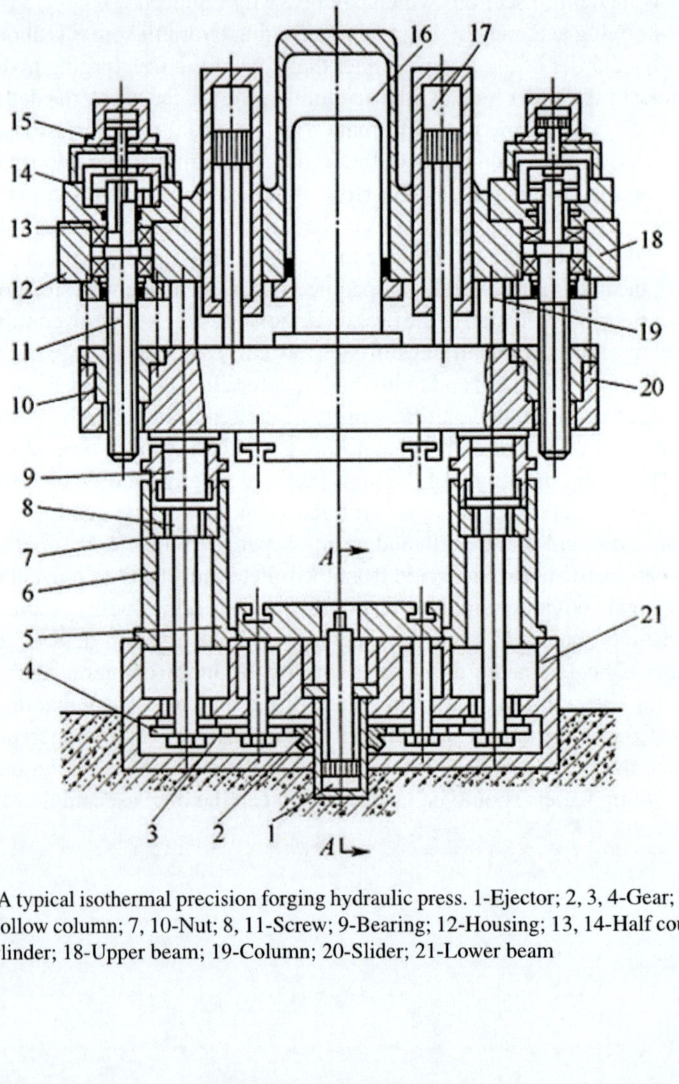

Fig. 7.5 A typical isothermal precision forging hydraulic press. 1-Ejector; 2, 3, 4-Gear; 5-Backing plate; 6-Hollow column; 7, 10-Nut; 8, 11-Screw; 9-Bearing; 12-Housing; 13, 14-Half coupling; 15, 16, 17-Cylinder; 18-Upper beam; 19-Column; 20-Slider; 21-Lower beam

the slider moves, the screw rotates in the nut. When the screw does not rotate, the slider stops. When the gear-type half coupling meshes with the half coupling that is connected to the screw, the screw stops moving. The half coupling cannot be rotated because it is located in the square groove of the housing.

In order to meet the requirements of isothermal precision forging, the velocity of the isothermal precision forging hydraulic press should be adjustable. The recommended adjustment range is 100 mm/s ~ 0.1 mm/s. The pressure can be maintained for a long time, that is, it can be maintained at the rated pressure for more than 30 min. In order to install dies, heating devices, cooling plates, heat insulation panels and other tooling, it is necessary to have a large shut height and sufficient work surface, preferably with a movable workbench. Meanwhile, an ejection device with sufficient ejection stroke and ejection force should be provided. In addition, the temperature control system is necessary to precisely control the forging temperature.

The isothermal precision forging hydraulic press is mainly used for precision forging of rib-web parts with large ratio of height and width. The forging force is almost linear with the increase of the projected area of the forging. When the height of the ribs increases and the thickness of the webs decreases, the forgings have a larger surface area per unit volume, which greatly increases the friction resistance and reduces the material temperature, resulting in an increase in the forging force. Therefore, the ratio of the surface area to the volume of the forging significantly affects the forging force. In addition, the direction of metal flow also has a great influence on the forging force sometimes.

Generally, the forging force can be estimated by Eq. (7.5). The average unit pressure p is 2 ~ 4 times the flow stress of the material. The larger value is used for closed precision forging, precision forging of thin web parts, and backward extrusion, and the smaller value is used for open die forging and forward extrusion.

7.3.4 New Medium and Small Precision Forging Hydraulic Press

In view of the narrow temperature range and strong rate sensitivity of high-strength aluminum alloy precision forging, Huazhong University of Science and Technology cooperated with Hubei Tri-ring Metalforming Equipment Co., Ltd. and Huangshi Huali Forging Machine Tool Co., Ltd. to develop new type of precision forging hydraulic press. At present, YK34J series single-action, multi-action and multi-directional precision forging hydraulic presses have been developed. The working principles, structural features and main technical parameters of these three precision forging hydraulic presses are introduced below.

1. Single-action precision forging hydraulic press

The traditional single-action hydraulic presses are mainly used for cold extrusion and closed cold precision forging. There are two outstanding problems, one is the poor guiding accuracy and rigidity of the frame, and the other is the low idle velocity and forging velocity.

Compared with traditional single-action hydraulic presses, the rigidity and guiding accuracy of the single-action precision forging hydraulic presses were greatly improved by replacing the traditional three-beam and four-pillar body with a frame-type integral body or a pre-stressed combined body, and replacing the four-pillar guide with an "X"-shaped precision guiding device and an extended slider. The hydraulic transmission system was installed on the top of the body to shorten the length of the pipeline system, which is beneficial to reduce the flow resistance of the pressure oil.

The developed YK34J-800 single-action precision forging hydraulic press is shown in Fig. 7.6. Its main technical parameters are listed in Table 7.2.

2. Multi-action precision forging hydraulic press

The structural diagram of the multi-action precision forging hydraulic press is shown in Fig. 7.7. The press consisted of hydraulic cylinder, side hydraulic cylinders, fast hydraulic cylinders, inner slider, outer slider, die floating cylinder, ejection

Fig. 7.6 YK34J-800 single-action precision forging hydraulic press

Table 7.2 Technical parameters of YK34J-800 precision forging hydraulic press

Project	Technical parameter		Value
Basic parameters	Nominal force/kN		8000
	Return force/kN		400
	Slider stroke/mm		500
	Maximum opening height/mm		1100
	Size of workbench /mm		1250 × 1000
Slider velocity	Idle velocity/(mm/s)		240
	Forging velocity/(mm/s)		13.1 ~ 20
	Return velocity/(mm/s)		130
Lower ejector	Ejection force/kN		200
	Ejection return force/kN		40
	Ejection stroke/mm		60
	Ejection velocity/(mm/s)		46
	Ejection return velocity/(mm/s)		115
Upper ejector	Ejection force/kN		100
	Ejection stroke/mm		30
Others	Main system pressure/MPa		28.5
	Motor power	Main motor/kW	55 × 2
		Auxiliary motor/kW	7.5
	Machine weight/kg		58,100

hydraulic cylinder and integral welded body. The piston rod of the main hydraulic cylinder was connected with the inner slider, and the plungers of the side hydraulic cylinders and the piston rods of the fast hydraulic cylinders were connected with the outer slider. The inner slider was completely contained in the outer slider, which has the characteristics of compact structure. The die floating cylinder and the ejection hydraulic cylinder were installed under the workbench. The upper ejector was installed in the center hole of the inner slider. The hydraulic drive system was composed of hydraulic components including low-pressure pump, high-pressure pump and high-pressure relief valve, electromagnetic reversing valve, one-way valve and overfill valve.

The developed Y28-800(400/400) multi-action precision forging hydraulic press is shown in Fig. 7.8. Its main technical parameters are listed in Table 7.3.

This hydraulic press was mainly used for closed precision forging of high-strength aluminum alloy parts. Before precision forging, the lower die of the separable die was installed on the workbench, while the upper die and the upper punch were installed under the outer and inner sliders, respectively. During precision forging, firstly, the heated billet was put into the lower die, and the upper die moved downward with

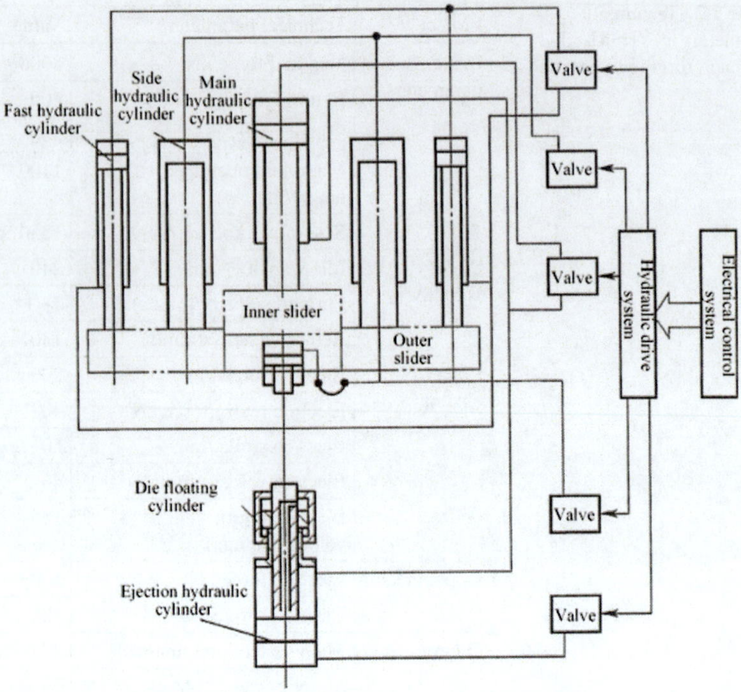

Fig. 7.7 Structural diagram of multi-action precision forging hydraulic press

the outer slider until it was closed with the lower die. Subsequently, the inner slider drived the upper punch down to exert force on the billet to fill the entire die cavity and form the required forging. It should be pointed out that the ejection cylinder could drive the lower punch to deform the billet synchronously or asynchronously with the upper punch. After the precision forging was completed, firstly, the inner slider drived the upper punch out of the forging and the die, and moved upward to return to the initial position. Then, the fast hydraulic cylinders drived the upper die to return to the initial position, and the piston rod of ejection cylinder pushed the forging out of the lower die through the lower punch or ejector rod in the lower die. The upper punch driven by the inner slider and the lower punch driven by the ejection cylinder could move synchronously at the same force and velocity in opposite directions, and also asynchronously at different forces and velocities. Therefore, the application range of the press is wide.

3. Multi-directional precision forging hydraulic press

The developed YK34J-1600/C1250 multi-directional precision forging hydraulic press is shown in Fig. 7.9. Its main technical parameters are listed in Table 7.4. The vertical frame body and the horizontal four-rod frame of the multi-directional precision forging hydraulic press employed a non-rigid connection. The middle part of the horizontal four-rod frame was supported on the workbench. With this structure,

Fig. 7.8 Y28-800 (400/400) multi-action precision forging hydraulic press

when the horizontal cylinder was loaded laterally, the elongation of the horizontal frame caused by the lateral force would not cause the vertical frame body to produce horizontal deformation, and ddi not affect the guiding accuracy of the slider.

This multi-directional precision forging hydraulic press could realize the following multi-directional die forging processes.

(1) The force generated by the main cylinder was used as the clamping force of the horizontally separable die, and the left and right horizontal cylinders performed synchronous or asynchronous lateral extrusion. The process was used for precision forging of equal-diameter two-way or different diameter two-way pipe joints.
(2) The force generated by the two horizontal cylinders on the left and right was used as the clamping force of the vertically separable die. The main cylinder drived the punch to complete the forging. The process was used for precision forging of double flanges and complex cylindrical parts with two or more annular grooves in the middle.
(3) The force generated by the main cylinder was the clamping force of the horizontally separable die. The left or right horizontal cylinder first performed

Table 7.3 Technical parameters of Y28-800 (400/400) multi-action extrusion hydraulic press

Technical parameter		Value
Slider	Nominal force of inner slider/kN	4000
	Nominal force of outer slider/kN	4000
	Stroke of inner slider/mm	600
	Stroke of outer slider/mm	600
Ejector	Ejection force/kN	2000
	Return force/kN	500
	Ejection stroke/mm	200
Stroke of side cylinder/mm		200
Shut height of inner slider/mm		500
Shut height of outer slider/mm		500
Size of workbench	From left to right/mm	1000
	From front to back/mm	1250
Size of inner slider	From left to right/mm	400
	From front to back/mm	400
Size of outer slider	From left to right/mm	1000
	From front to back/mm	1250
Velocity of inner slider	Idle velocity/(mm/s)	150
	Forging velocity/(mm/s)	20 ~ 40
	Return velocity/(mm/s)	150
Velocity of outer slider	Idle velocity/(mm/s)	150
	Forging velocity/(mm/s)	20 ~ 40
	Return velocity/(mm/s)	15
Velocity of ejector	Ejection velocity/(mm/s)	40 (adjustable)
	Return velocity/(mm/s)	150

Fig. 7.9 YK34J-1600/C1250 multi-directional precision forging hydraulic press

pre-forging, and the other cylinder completed the final forging. It was used for two-step precision forging of complex forgings.

7.3.5 *Servo Hydraulic Press*

In view of the high rate sensitivity of aluminum alloys, it is also an advanced solution for aluminum alloy precision forging to use a servo-hydraulic press with high and variable velocity within a certain range. In response to this demand, a Y68SK high-performance precision forging servo hydraulic press was developed. The "pump control servo" technology which realizes precise control of flow and pressure through control of the torque and velocity of the pump was employed. The hydraulic system was composed of electro-hydraulic servo drives, three-phase AC permanent magnet synchronous motors, high-performance dedicated servo pumps, and pressure sensors. Since the output flow of the oil pump was proportional to the motor velocity, and the pressure in the oil circuit was proportional to the output torque of the motor, the precise control of the torque and speed of the pump could be realized by using the vector, field weakening and special PID control algorithm to control the system pressure and flow with double closed-loop. The servo press realized accurate supply in terms of the actual required flow and pressure, which could eliminate energy loss and overcome the problem of the "valve-controlled servo" system, that is, the oil temperature rises too quickly. The highest energy saving rate was 70%, and the average energy saving rate was 30%. The developed Y68SK-315 servo hydraulic press is shown in Fig. 7.10. The main technical parameters of the series products are listed in Table 7.5.

Table 7.4 Technical parameters of YK34J-1600/C1250 multi-directional precision forging hydraulic press

Project	Technical parameter	Value
Main slider	Nominal force/kN	16,000
	Return force/kN	700
	Slider stroke/mm	600
	Maximum opening height/mm	1300
	Idle velocity/(mm/s)	250
	Forging velocity/(mm/s)	7.2 ~ 30
	Return velocity/(mm/s)	250
Horizontal sliders	Nominal force/kN	12,500
	Return force/kN	500
	Slider stroke/mm	300
	Opening width/mm	1000
	Idle velocity/(mm/s)	150
	Forging velocity/(mm/s)	7.5 ~ 30 (adjustable in 4 levels)
	Return velocity(mm/s)	200
	Synchronization error of the left and right sliders/mm	≤ 0.2
Lower ejector	Ejection force/kN	500
	Return force/kN	80
	Maximum ejection stroke/mm	200
	Ejection velocity/(mm/s)	50
	Return velocity/(mm/s)	120
Others	Working pressure of main system /MPa	28.5
	Size of workbench/mm × mm	1500 × 1000

Compared with traditional die forging hydraulic press, Y68SK servo hydraulic press has several characteristics, including high flexibility, high efficiency, low noise, low energy consumption and convenient maintenance.

The slider movement curve can be set according to different processes to realize the “free movement” of the slider. The slider can be set to a free working mode or a working mode of bottom dead center pressure keeping, which greatly improves the application scope of the hydraulic press. The number of strokes of the slider can be set in a larger range according to the needs of the actual situation, and the velocity of the slider can be adjusted in a wide range, which is beneficial to improve production efficiency. The servo hydraulic press can reduce the noise by 3 ~ 10 dB during work, and can reduce the noise by more than 30 dB when the movement of slider stops, which greatly reduces the impact on the operator and the environment. Compared with the traditional hydraulic press, the servo hydraulic press can save

Fig. 7.10 Y68SK-315 servo hydraulic press

Table 7.5 Main technical parameters of Y68SK series precision forging servo hydraulic press

Model		Y68SK-315	Y68SK-400	Y68SK-500	Y68SK-630
Nominal force/kN		3150	4000	5000	6300
Opening size/mm		750	1000	1100	1300
Stroke/mm		400	500	600	800
Size of workbench From left to right × From front to back/(mm × mm)		700 × 800	1000 × 1000	1000 × 1000	1000 × 1200
		1000 × 1000	1000 × 1200	1000 × 1200	1200 × 1400
		1000 × 1200	1200 × 1400	1200 × 1400	1600 × 1800
Velocity of slider /(mm/s)	Idle velocity	400	400	400	400
	Velocity of 50% load	50	50	50	50
	Velocity of 100% load	35	30	30	25
	Return velocity	300	300	300	300

energy consumption of 30% to 70%. The specific data varies according to the precision forging process and the production cycle. Since the servo proportional hydraulic valve, velocity control circuit, and pressure control circuit in the hydraulic system are eliminated, the hydraulic system is greatly simplified. The cleanliness requirements of hydraulic oil are far less than that of servo proportional hydraulic systems, which reduces the influence of hydraulic oil pollution on the system.

7.3.6 Large Hydraulic Press

Large hydraulic presses are mainly used to form large die forgings of aluminum alloy, magnesium alloy and titanium alloy in the aviation industry. On the one hand, the horizontal projection area of aeronautical components is large, some up to several m^2. On the other hand, the high-strength metal materials are often employ for the components. For example, the unit pressure of aluminum alloy and magnesium alloy is about 200 ~ 800 MPa, the unit pressure of titanium alloy can reach more than 1000 MPa, therefore, the large hydraulic press is required. At present, the world's largest hydraulic press is the 800MN die forging hydraulic press (Fig. 7.11), which was built in 2013 by China Second Heavy Machinery Group Corporation.

The main machine of the hydraulic press consisted of C-shaped frame, five main working hydraulic cylinders, vertical perforation system, upper beam, combined movable beam, combined lower beam, movable workbench, four return hydraulic cylinders, and four synchronous cylinders. The pre-stressed combined body used more than 60 tie rods with diameters of Φ160mm ~ 900 mm. The working oil pressure of the five main working cylinders was 0 ~ 63 MPa, and the maximum working pressure of a single cylinder was 160MN. The total force of the five cylinders was 800 MN. The lifting force of the single cylinder of the workbench was 2400kN, and the total lifting force was 19200kN.

The movable beam of the hydraulic press was composed of two middle beams, two side beams, four guide rods, two upper pads, middle pads, pillars, and lower plates. The two middle beams were tensioned by 10 tie rods with a diameter of Φ450mm, and the pre-tightening force applied on each tie rod was 22,000kN. The two side beams were locked by two Φ550mm tie rods and the middle beam. The pre-tightening force applied on each tie rod was 2,800kN. The stroke of the movable beam was 2 m. The dimension the workbench was 8 m × 4 m, and the maximum space height for installing die was 4.5 m.

The main parameters such as pressure, velocity, time, and position could be monitored by the control system of the hydraulic press. In order to meet the requirements of a variety of die forging processes, three loading and movement modes were designed: (1) the pressure is 400MN, and the velocity is 60 mm/s; (2) the pressure is 600MN, and the velocity is 40 mm/s; (3) the pressure is 800MN, and the velocity is 30 mm/s. Within a certain mode, the pressure and velocity could be infinitely adjusted. The reasonable configuration of forging force and forging velocity was conducive

Fig. 7.11 The 800MN hydraulic press

to adapting to the process requirements of the forgings of different materials and different structures.

The nominal force of the hydraulic press is 800 MN, which can realize the overall manufacture of most large aerospace die forgings, replacing the manufacturing method of segmented forging and welding, which is not only conducive to improving material utilization and production efficiency, but also improving the quality of forgings.

7.4 Screw Press

7.4.1 Screw Presses for Aluminium Alloy Forging

At present, the most used screw presses for aluminium alloy forging are mainly divided into clutch type screw presses, non-direct drive type electric screw presses and direct drive type electric screw presses. The difference between the clutch screw press and the traditional screw press is not in the transmission mode (such as friction, hydraulic pressure, direct motor transmission), but in the different working mode of the flywheel. The flywheel of the clutch screw press is driven by the main motor through the V-belt to make it rotate freely in one direction. When working, the clutch piston is pushed by hydraulic pressure to combine the clutch driven disc with the flywheel. Since the screw is integrated with the clutch driven disc, the screw is driven to rotate. Through the rotation of the nut that is connected with the slider, the slider is driven to move down to achieve forging. When the velocity of the flywheel decreases to a certain value, the disengagement mechanism of the clutch system disengages the clutch by controlling the ejector rod to open the hydraulic control valve. Meanwhile, the slider returns upward under the drive of the return cylinder [2].

The working principle of the non-direct drive electric screw press is that the motor drives the flywheel and the screw which connected with the flywheel through a pinion gear, and then the rotating screw drives the slider to reciprocate through the nut that is fastened with the slider to realize forging [3, 4]. When the motor reaches the velocity required by the striking energy, the energy stored in the flywheel is used for forging. After the energy of the flywheel is released, the motor immediately drives the flywheel to reverse. After a certain angle of rotation, the motor switches to the braking state and the slider returns to the initial position.

Based on this principle, a series of small and medium-sized electric screw presses with different structures were developed. Among them, the main technical parameters of a series of J58K electric screw presses developed by Wuhan Newish Technology Co., Ltd. are listed in Table 7.6. The J58K-2500 electric screw press is shown in Fig. 7.12.

Compared with the traditional friction press, the electric screw press has the following characteristics.

(1) The transmission chain was shorter. Direct gear transmission was used to replace belt and friction transmission, and the pinion gear was made of a special non-metallic material with better self-lubricating function. The contact friction coefficient was greatly reduced, so the moment of inertia was small and the transmission efficiency was greatly improved. The energy consumption could be saved by more than 30%, and the raito of energy saving increased significantly with the increase of the press tonnage. Based on the transmission chain, the blow energy of the press could be accurately controlled, which is beneficial to improve the service life of the forging die and the press.

(2) The lengthened slider and lengthened guiding device was employed to improve the anti-eccentric capacity, which is not only conducive to improving the dimensional accuracy of forgings, but also conducive to the realization of multi-station die forging.

(3) It was conducive to integration with robots and other equipment to establish an automated production line.

Table 7.6 Main technical parameters of J58K electric screw press

Model	J58K-160	J58K-315	J58K-630	J58K-1000	J58K-2500	J58K-4000	J58K-6300
Nominal force/kN	1600	3150	6300	10,000	25,000	40,000	63,000
Allowable force for long-term operation/kN	2500	5000	10,000	16,000	40,000	63,000	100,000
Energy/kJ	10	20	80	160	500	1000	1650
Slider stroke/mm	300	380	450	500	650	750	850
Number of trips/ (times/min)	30	26	20	18	14	11	8
Minimum shut height/mm	500	550	720	750	1050	1460	1700
Workbench size (from left to right/from front to back)/mm	600/560	700/640	820/900	920/1050	1250/1400	1640/2000	2100/2000

7.4.2 Direct Drive Electric Screw Press

Direct drive electric screw press is the latest development of electric screw press. Its structure consists of cooling motor, main motor stator, rotor, screw, upper beam, thrust bearing and tie rods. The main motor stator is fixed on the upper beam, the main motor rotor is fixed on the outer cylindrical surface of the flywheel. The flywheel is connected with the screw as a whole. And the rest is the same as the non-direct drive electric screw press. When the main motor is energized in the forward and reverse directions, the rotor rotates rapidly in the forward and reverse directions under the action of the magnetic field and drives the screw to rotate at the same velocity. In this way, the slider is driven to reciprocate downward and upward to realize forging.

The developed J58ZK direct-drive electric screw press with a nominal force of 5MN is shown in Fig. 7.13. Its main technical parameters are listed in Table 7.7.

Compared with the non-direct drive electric screw press, the direct drive electric screw press has a shorter transmission chain, so the transmission efficiency is higher, and the energy saving effect is better. Meanwhile, the number of slide strokes and production efficiency are also higher.

Fig. 7.12 J58K-2500 electric screw press

7.4.3 Calculation of the Nominal Force of Screw Press

The nominal force (P_g) of the screw press can be determined according to the following empirical equation.

$$P_g = F/q \tag{7.7}$$

where F is the force required for deformation of the forging, q is the coefficient. For the forging with small deformation stroke, q takes 1.6; for the forging with slightly larger deformation stroke, takes 1.3; for the forging with large deformation stroke and large deformation energy, takes 0.9 ~ 1.1.

F can be determined according to the following empirical equation.

$$F = K_s A \tag{7.8}$$

where A is the horizontal projected area of the forging, K_s is the coefficient. When the outline of aluminum alloy forgings are required to be clear, K_s takes 500 MPa;

Fig. 7.13 J58ZK-500 direct drive electric screw press

Table 7.7 Main technical parameters of J58ZK-500 direct drive electric screw press

Parameter	Value
Nominal force/MN	5
Allowable force for long-term operation/MN	8
Energy/kJ	50
Slider stroke/mm	400
Number of trips/(times/min)	22
Minimum shut height/mm	450
Size of workbench (from left to right × from front to back)/(mm × mm)	750 × 730
Motor power/kW	80

when the forgings have rounded corners and smooth outline, takes 300 MPa; for forgings with very small thickness, takes 800 MPa.

In addition, the following equation can also be used to calculate the nominal force of the press.

$$P_g = a\left(2 + 0.1\frac{A\sqrt{A}}{V}\right)\sigma_b A \tag{7.9}$$

where V is the forging volume, σ_b is the tensile strength of the material, and a is the coefficient. For die forging without flash in the closed integral die, a takes 3; for die forging without flash in the closed separable die, a takes 5; for extrusion, a takes 5.

References

1. Zhang ZW (1988) Forging technology. Machinery Industry Press, Beijing
2. Yang YH, Liu D, Luo ZJ (2009) Working characteristics and numerical simulation of forging process for clutch screw press. Acta Aeronautica et Astronautica Sinica 30(7):1346–1357
3. Xiong XH, Li JC, Feng Y, Huang SH (2007) CNC electrical screw presses. Forging & Stamping Technology 32(5):110–113
4. Feng Y, Huang SH, Li JC, Xiong XH (2009) Electric screw press driven by permanent magnet synchronous motor. Forging & Stamping Technology 34(6):112–116